AF804326

Lecture Notes on Multidisciplinary Industrial Engineering

Series Editor

J. Paulo Davim, Department of Mechanical Engineering, University of Aveiro, Aveiro, Portugal

"Lecture Notes on Multidisciplinary Industrial Engineering" publishes special volumes of conferences, workshops and symposia in interdisciplinary topics of interest. Disciplines such as materials science, nanosciences, sustainability science, management sciences, computational sciences, mechanical engineering, industrial engineering, manufacturing, mechatronics, electrical engineering, environmental and civil engineering, chemical engineering, systems engineering and biomedical engineering are covered. Selected and peer-reviewed papers from events in these fields can be considered for publication in this series.

More information about this series at http://www.springer.com/series/15734

V. Sivasubramanian · S. Subramanian
Editors

Global Challenges in Energy and Environment

Select Proceedings of ICEE 2018

Springer

Editors
V. Sivasubramanian
National Institute of Technology Calicut
Kozhikode, Kerala, India

S. Subramanian
Indian Institute of Science
Bengaluru, Karnataka, India

ISSN 2522-5022 ISSN 2522-5030 (electronic)
Lecture Notes on Multidisciplinary Industrial Engineering
ISBN 978-981-13-9212-2 ISBN 978-981-13-9213-9 (eBook)
https://doi.org/10.1007/978-981-13-9213-9

© Springer Nature Singapore Pte Ltd. 2020
This work is subject to copyright. All rights are reserved by the Publisher, whether the whole or part of the material is concerned, specifically the rights of translation, reprinting, reuse of illustrations, recitation, broadcasting, reproduction on microfilms or in any other physical way, and transmission or information storage and retrieval, electronic adaptation, computer software, or by similar or dissimilar methodology now known or hereafter developed.
The use of general descriptive names, registered names, trademarks, service marks, etc. in this publication does not imply, even in the absence of a specific statement, that such names are exempt from the relevant protective laws and regulations and therefore free for general use.
The publisher, the authors and the editors are safe to assume that the advice and information in this book are believed to be true and accurate at the date of publication. Neither the publisher nor the authors or the editors give a warranty, expressed or implied, with respect to the material contained herein or for any errors or omissions that may have been made. The publisher remains neutral with regard to jurisdictional claims in published maps and institutional affiliations.

This Springer imprint is published by the registered company Springer Nature Singapore Pte Ltd.
The registered company address is: 152 Beach Road, #21-01/04 Gateway East, Singapore 189721, Singapore

Preface

Global Challenges in Energy and Environment is a proceedings book of the First International Conference on Energy and Environment: Global Challenges (ICEE 2018) held in the Department of Chemical Engineering, NIT Calicut, during 9 and 10 March 2018. National Institute of Technology Calicut (NITC) is one of the premier technical institutions in southern India located in Calicut which is a metropolitan city in the state of Kerala on the Malabar Coast.

Industrial growth, economic development, consumerization and urbanization have resulted in problems of environmental pollution. Environmental engineering is the integration of science and engineering principles to improve the natural environment to provide healthy water, air and land for human habitation and for other organisms and to remediate pollution sites. Similarly, energy is the source of economic growth. Energy consumption reflects the state of the development of the country. Energy engineering is one of the recent engineering disciplines to emerge. It is increasingly seen as a major step forward in meeting carbon reduction targets. In the past two decades, the development in these fields yielded various attractive results that are beneficial for the development of humanity.

This book focuses on the state-of-the-art technologies pertaining to energy and environmental research. This book provides a platform for experts from India and abroad to promote association and knowledge transfer by promising interaction within the wider research community in the fields of chemical and biochemical engineering, wastewater treatment and energy and environmental sustainability using green technologies and industrial applications.

Major areas covered include advanced chemical processes, air pollution and control, biomass conversion, energy policy, planning and management, integrated energy systems, membrane engineering, new perspectives in renewable energy, novel separation processes, nuclear energy applications, photo- and electrochemical engineering, polymeric materials, solid waste management and wastewater treatment.

We expect that this book will be a trigger for further related research and technology improvements in energy and environment.

Kozhikode, India V. Sivasubramanian
Bengaluru, India S. Subramanian

Acknowledgements As a token of our appreciation, it is our sincere pleasure to acknowledge Ministry of Human Resource Development (MHRD), National Institute of Technology Calicut (NITC), Department of Science and Technology (DST), Department of Biotechnology (DBT), Kerala State Council for Science, Technology and Environment (KSCSTE), Hindustan Petroleum Corporation Limited (HPCL), VKC Pride and Sinsil International Private Limited, Karnataka, who supported us technically and financially.

Contents

About the Editors

Dr. V. Sivasubramanian is a Professor and Former Head of the Department of Chemical Engineering at NIT Calicut, Kozhikode. After completing his Masters in Petroleum Refining and Petrochemicals, he obtained his M.B.A. from IGNOU and his Ph.D. from Anna University. He has worked as Senior Plant Engineer in Tamil Nadu Industrial Explosives Limited. His research interests include fluidization engineering, wastewater treatment, biofuels, solid waste management and electro-coagulation. He has published several research articles in international journals and conference proceedings. He is a Fellow Member of FICCE and a Life Member of IIChE and GEPRA.

Dr. S. Subramanian is a Professor at the Department of Materials Engineering, Indian Institute of Science, Bangalore. He has a Ph.D. in Metallurgical Engineering from Mysore University. He worked as a postdoctoral researcher at the University of British Columbia, Vancouver, Canada and was a visiting scientist at the Lulea University of Technology, Lulea, Sweden. His research areas include mineral processing and environmental biotechnology. Dr. Subramanian is a Fellow Member of the Indian Institute of Metals and the Geological Society of India. He is a Life Member of the Electrochemical Society of India and the Indian Institute of Mineral Engineers. He is also the present Secretary for Karnataka State Council for Science and Technology.

Chapter 1
Single-Step Electrodeposition of CZTS Thin Film for Solar Cell Application: Effect of Annealing Time

Akanksha Paraye, Arpita Shukla, R. Manivannan and S. Noyel Victoria

Abstract Silicon-based solar cells are widely used solar cells, but it needs a very thick absorber layer. Absorber layers based on cadmium telluride (CdTe), copper indium gallium diselenide (CIGS), copper indium diselenide (CIS), and copper zinc tin sulfide (CZTS) are extensively studied for thin-film solar cells. The use of CdTe for thin-film solar cells is less preferred due to its toxicity. In comparison with CIGS, thin-film solar cells with CZTS absorber layers are preferred since the raw materials are earth-abundant, nontoxic, and inexpensive. The CZTS materials have a high absorption coefficient of 10^4 cm^{-1} and have the band gap energy of 1.4–1.5 eV. In this work, single-step electrodeposition of Cu_2ZnSnS_4 (CZTS) thin-film on gold-coated slide using glycine as a complexing agent was carried out. The effect of different annealing time on surface morphology, crystallite size, and its elemental composition was studied. X-ray diffraction (XRD) analysis reveals the kesterite phase of CZTS. Crystallite size increases with increase in the annealing time. Deposited CZTS thin film annealed for 30 min shows the elemental composition near to the desired stoichiometry (Cu:Zn:Sn:S = 2:1:1:4).

Keywords Thin film · Electrodeposition · CZTS · Annealing time · XRD · Crystallite size

1.1 Introduction

The enhancement of the world population, the consumption of conventional energy, and its harmful side effects on environment are also increased. It is expected that demand of energy to be twice that of today's energy demand, i.e., approximately 30 TW. The consumption and demand of this energy for next few years will be very challenging [1]. The present energy demand is fulfilled by fossil fuel such as oil,

A. Paraye · A. Shukla · R. Manivannan · S. Noyel Victoria (✉)
Department of Chemical Engineering, National Institute of Technology Raipur, Raipur, Chhattisgarh 492010, India
e-mail: snvictoria.che@nitrr.ac.in

© Springer Nature Singapore Pte Ltd. 2020

V. Sivasubramanian and S. Subramanian (eds.), *Global Challenges in Energy and Environment*, Lecture Notes on Multidisciplinary Industrial Engineering, https://doi.org/10.1007/978-981-13-9213-9_1

coal, and gasses which emits toxic gasses after its consumption, and on the other hand, they are not promising, since their quantity will be not sufficient to fulfill the future energy demand. Therefore, the world is focusing on renewable energy sources to fulfill this energy demand with the least side effect on environment. Solar energy is found to be an alternative and promising way to fulfill the present and future energy demand, and it is also known as one of the green and clean energy. One of the best applications of solar energy is solar cell or photovoltaic cell. Solar cell is an electric device which converts the solar energy directly into the electricity due to photovoltaic effect which is a physical and chemical phenomenon. Solar cell can be broadly divided into three generations. The first generation solar cells are also known as a conventional, traditional, or wafer-based cell which is made of crystalline silicon. Silicon-based solar cell is widely in use. But there are many demerits of silicon—such as it has an indirect band gap material, with low absorption coefficient, it needs thick absorber layer, and it losses its efficiency at higher temperature, i.e., in hot sunny days. Second-generation solar cells are basically thin-film solar cells, which include amorphous silicon; cadmium telluride (CdTe); and copper indium gallium diselenide (CIGS). Materials used in second-generation solar cells as an absorber material are rare, expensive, and toxic. The availability of indium (In) and selenium (Se) makes them expensive, whereas cadmium (Cd) is toxic to environment [2]. Third-generation solar cells include a number of thin-film technologies often described as emerging photovoltaic. These solar cells are made of organic-based materials. Most of them have not yet been commercially applied and are still in the research or development phase. These solar cells are not promising cells as they loss their efficiency with very short time period. In last ten years, thin-film solar cells have attracted the researchers most due to its requirement of less absorber material which reduces the cost of thin films; absorber materials for thin film can be flexibly deposited on substrates such as glass, stainless steel, and plastic, especially suitable for solar building integration [1]. Inorganic absorber materials presently in use are copper indium gallium diselenide (CIGS), copper indium diselenide (CIS), and cadmium telluride (CdTe). Elements of these absorber materials also have some demerits, such as availability of indium (In) and selenium (Se) which makes them expensive and cadmium (Cd) which is toxic [3]. Absorber materials for thin-film technology with desirable requirements such as being inexpensive and environmental-friendly with direct band gap are under study. Compared to commercialized absorber materials, CZTS have the benchmark characteristics such as it is composed of only earth abundant, inexpensive, and non-toxic elements. Cu_2ZnSnS_4 (CZTS) is a promising absorber material, with direct band gap of 1.4–1.7 eV and high absorbance coefficient of 10^4 cm^{-1} for thin-film solar cell, which is expected to become the ideal absorber layer material for next-generation thin-film solar cell [4]. There are various routes for the preparation of CZTS nanoparticles and thin films for the solar cell application. Techniques such as thermal evaporation [5], atom beam sputtering [6], pulsed laser deposition [7], hybrid sputtering [8], etc., are well-established routes for the deposition of CZTS thin films. But on the other hand, they are very expensive and require high energy to operate, with very high vacuum pressure. However, there are other techniques also which demands low energy such as successive ionic layer adsorption and reaction (SILAR)

[9], chemical bath deposition [10], spray pyrolysis [11], and electrochemical deposition [12]. Electrochemical deposition is one of the well-known low energy demand technique as it provides scalable deposition of the desired film at room temperature. In this present work, the electrodeposition of CZTS thin film on gold-coated substrate under potentiostatic mode was carried out. Glycine was used as a complexing agent in this study. The deposited sample was then annealed for different time periods at constant temperature. The effect of annealing time on its surface morphology, crystallite size, and its elemental composition was studied.

1.2 Experimental

Gold-coated microscopic slide having thickness of 100 Å (Sigma-Aldrich, 99.99% Au) was used as a substrate for electrodeposition of CZTS. Precursors of analytical grades (Merck) were used for the deposition of CZTS thin film. Copper sulfate pentahydrate ($CuSO_4{\cdot}5H_2O$) was used as a copper precursor, zinc sulfate heptahydrate ($ZnSO_4{\cdot}7H_2O$) was used as a zinc precursor, tin chloride ($SnCl_2$) was used as a tin precursor, and sodium thiosulfate ($Na_2S_2O_3$) was used as a sulfur precursor. Glycine as a complexing agent was used in this study to narrow down the differences in the cathodic potential. Electrolyte was prepared by using deionized water. The electrolyte contained 10, 20, 5, 80, and 100 mM of Cu, Zn, Sn, S, and glycine, respectively. HNO_3 or KOH solutions were used to adjust the pH at 2.5. CZTS deposition was performed under potentiostatic mode.

Deposition studies of CZTS were performed in an electrochemical workstation, CHI 660E, CH Instruments, USA. Experiment for CZTS deposition was carried out using three electrode configurations (counter, reference, and working). Platinum wire was used as the counter electrode, and Ag/AgCl was used as the reference. The working electrode was prepared by using gold-coated microscopic slide, having dimensions of 10 mm × 10 mm × 1.1 mm. Cyclic voltammetry (CV) runs for the electrolyte having all the precursors with and without complexing agent were carried out to find out the suitable deposition potential of CZTS. CV runs were carried out with scan rate of 0.5 v/sec. Deposition of CZTS was done for 20 min under potentiostatic mode. The obtained deposition was then washed using deionized water, dried, and annealed at 300 °C for different time periods 30, 45, 60, and 75 min in inert environment. Scanning electron microscope (SEM) (Zeiss Evo-Model EVO 18) was used to study the morphology of the deposits. Energy-dispersive X-ray analysis (EDX) (INCA 250 EDS with X-MAX 20 mm detector) was also done for elemental composition. X-ray diffraction (XRD) (PANalytical 3 kW X'pert) was used to analyze the crystalline nature of the deposits at different annealing conditions. The energy band gap was calculated by using absorption spectra obtained by Shimadzu UV-1800, UV–visible spectrophotometer (UV–vis).

1.3 Results and Discussion

1.3.1 Cyclic Voltammetry

The cyclic voltagrams obtained for CZTS in the presence and absence of complexing agent are shown in Fig. 1.1. Elements such as copper, zinc, tin, and sulfide have different deposition/reduction potentials, due to which it is very hard to co-electrodeposit from a single electrolyte [13]. In order to narrow down this reduction potential gap between all the elements, glycine was used as a complexing agent to an electrolyte and reduction potential was optimized [14]. The effect of complexing agent can be clearly seen by obtained cyclic voltagrams shown in Fig. 1.1. In the absence of glycine, the cathodic sweep presents two peaks. After addition of glycine, both the reduction potential shifted to the negative direction and a well-defined peak was observed at −0.86 V (Vs. Ag/AgCl). Moreover, the current density increases with addition of glycine which is due to the complexing action of the same.

1.3.2 Scanning Electron Microscopy

In order to analyze the effect of annealing time on surface morphology and elemental composition (Table 1.1) of the CZTS thin films, SEM measurement was carried out. Figure 1.2 shows the SEM images of the deposits annealed for different durations at 300 °C. Morphological study for all the samples annealed for different time duration shows agglomeration of particles in nature, and this helps in preventing recombination which is one of the desirable characteristics for solar cell applications [14, 15]. Sample annealed for 30 min shows small and spherical particles in their shape. Sample annealed for 45 min showed bigger-sized agglomerates compared to sample annealed for 30 min. Similarly, samples annealed for 60 min show agglomerates formed out of flaky particles. Sample synthesized for 60 min shows the formation of

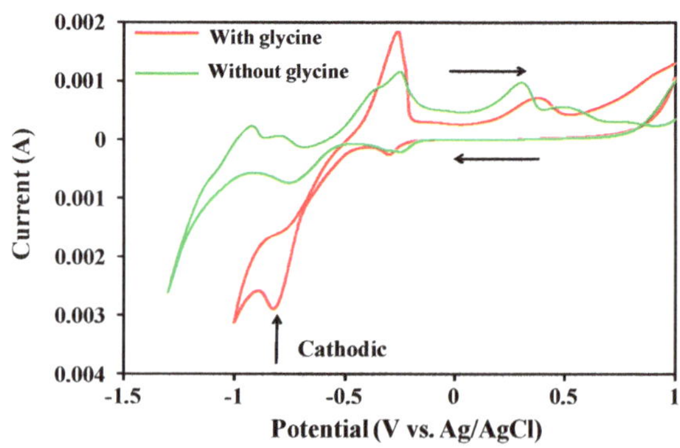

Fig. 1.1 Cyclic voltagrams in the presence and absence of complexing agent

Table 1.1 Effect of annealing time on elemental composition (atomic %)

Element	30 min	45 min	60 min	75 min
Cu	35.9	40.4	28.6	27.7
Zn	7.9	9.3	14.5	12.9
Sn	17.1	11.3	19.1	23.7
S	39.1	38.9	37.9	35.6
S/Cu	1.1	1.0	1.3	1.3
Cu/(Sn + Zn)	1.4	1.9	0.9	0.8

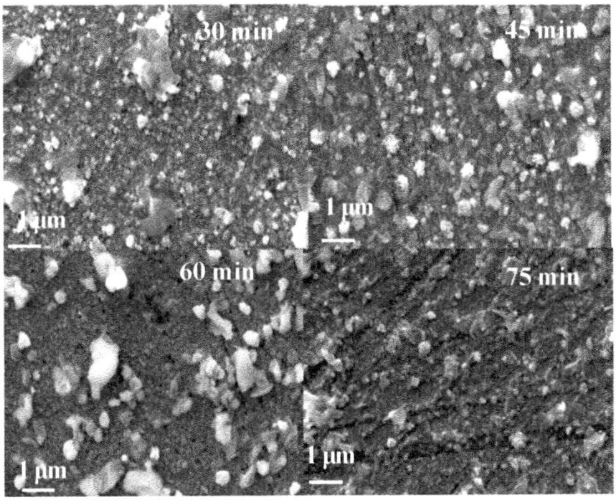

Fig. 1.2 Effect of annealing time on surface morphology

cracks. The samples annealed for 75 min show the presence of lesser agglomerates when compared to other samples.

The effect of different annealing time on elemental composition was also carried out. It can be seen from Table 1.1 that the sulfur concentration is affected to a lesser extent with annealing time. The samples annealed for 60 and 75 min are copper poor. Further, optimization is necessary to achieve desirable stoichiometry.

1.3.3 X-ray Diffraction

The X-ray diffraction patterns of annealed samples for different annealing time are shown in Fig. 1.3. The XRD spectra recorded for sample annealed at 300 °C for 30 and 45 min show the amorphous nature of the deposits. Small diffraction peaks were observed at 28.6° which correspond to (1 1 2) crystal plane of kesterite CZTS [COD (Crystallography Open Database) 96-900-4751]. It is observed that with increase in annealing time, the intensity (1 1 2) of diffraction peak becomes sharp, which indicates that the crystalline nature of CZTS thin film is also improved with increase

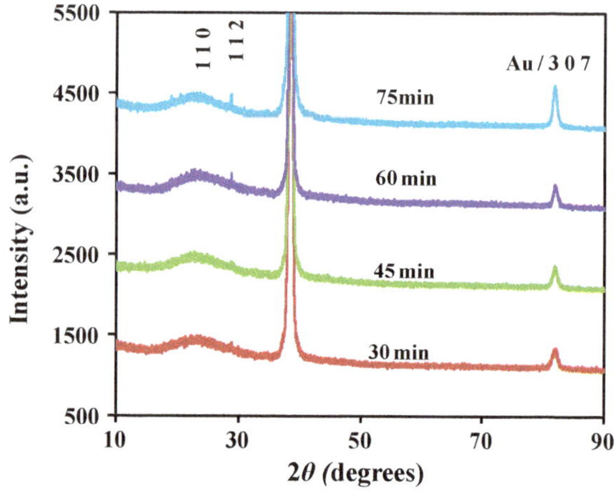

Fig. 1.3 X-ray diffraction patterns of CZTS thin films annealed for different annealing time

in annealing time. The diffraction peak positions are not shifted with annealing time, suggesting that CZTS phase is stable and its formation is independent of annealing time. XRD spectra obtained for all the samples also show that no secondary phases are formed.

1.3.4 UV–Visible Spectroscopy

The absorbance spectra for sample annealed for 30 min are shown in Fig. 1.4. It was found that the sample shows a good absorbance in the entire visible spectrum, thus suitable for solar cell applications. The band gap of the sample was calculated from absorbance data by using the Tauc relation [14]

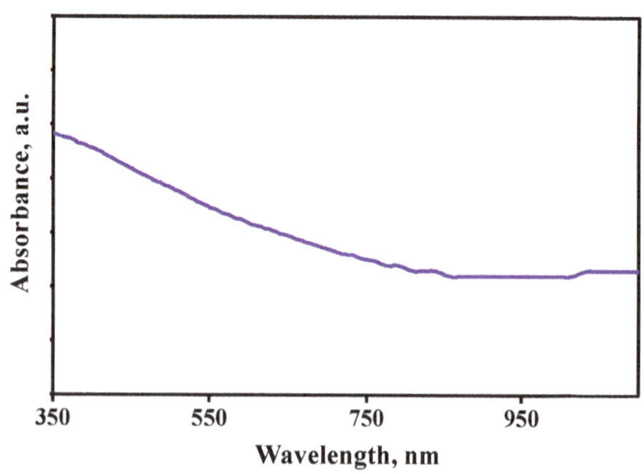

Fig. 1.4 Absorbance spectra of deposited CZTS annealed at 300 °C for 30 min

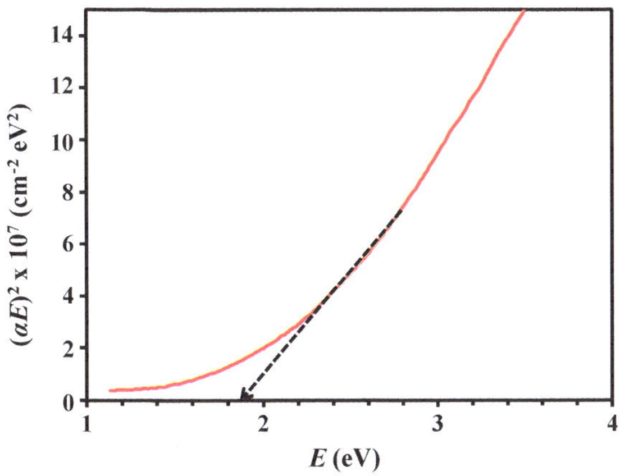

Fig. 1.5 Tauc plot of film annealed at 300 °C for 30 min

$$\alpha = \frac{A(E - E_g)^n}{E}$$

where

- α absorption coefficient.
- E photon energy in eV.
- E_g band gap energy in eV.
- n constant which is assigned values 1/2, 3/2, and 3 for direct allowed, indirect forbidden, and indirect allowed materials, respectively.

The Tauc plot for sample annealed at 300 °C for 30 min is shown in Fig. 1.5. It is observed that the band gap for the sample is 1.85 eV, which is an optimal band gap for the solar cell applications [1–12].

1.4 Conclusion

The effect of different annealing time for the CZTS deposits formed out of single-step electrodeposition was studied. The SEM results show the presence of agglomerates for the samples annealed for 30, 45, and 60 min. The deposits annealed for 60 min showed the presence of cracks. XRD result shows that with increase in annealing time, crystallinity improved slightly. The UV–visible spectroscopy studies show that the deposits exhibit good absorbance in the visible spectrum. The band gap of the particles was 1.85 eV.

Acknowledgements Authors would like to thank the Department of Science and Technology—Solar Energy Research Initiative (DST-SERI) for financing the research work under the Grant No. SEDST/TMC/SERI/2K12/57 and DST-SERB under the Grant No. SR/FTP/ETA-412/2013 for the analytical facilities.

References

1. Song, X., Ji, X., Li, M., Lin, W., Luo, X., Zhang, H.: A review on development prospect of CZTS based thin film solar cells. Int. J. Photoenergy **613173**, 1–11 (2014)
2. Bagher, A.M., Vahid, M.M.A., Mohsen, M.: Types of solar cells and application. Am. J. Optics Photonics **3**(5), 94–113 (2015)
3. Green, M.A., Emery, K., Hishikawa, Y., Warta, W., Dunlop, E.D.: Solar cell efficiency tables (Version 45). Prog. Photovoltaics Res. Appl. **23**(1), 1–9 (2015)
4. Kumar, M., Dubey, A., Adhikari, N., Venkatesan, S., Qiao, Q.: Strategic review of secondary phases, defects and defect-complexes in kesterite CZTS–Se solar cells. Energy Environ. Sci. **8**(11), 3134–3159 (2015)
5. Wang, K., Gunawan, O., Todorov, T., Shin, B., Chey, S.J., Bojarczuk, N.A., Mitzi, D., Guha, S.: Thermally evaporated Cu_2 $ZnSnS_4$ solar cells. Appl. Phys. Lett. **97**(14) 143508-1-3 (2010)
6. Katagiri, H.: Survey of development of CZTS-based thin film solar cells. In: Photonics (ICP), IEEE 3rd International Conference, pp. 345–349 (2012)
7. Moholkar, A.V., Shinde, S.S., Babar, A.R., Sim, K.U., Kwon, Y.B., Rajpure, K.Y., Kim, J.H.: Development of CZTS thin films solar cells by pulsed laser deposition: influence of pulse repetition rate. Sol. Energy **85**(7), 1354–1363 (2011)
8. Tanaka, T., Nagatomo, T., Kawasaki, D., Nishio, M., Guo, Q., Wakahara, A., Ogawa, H.: Preparation of Cu_2ZnSnS_4 thin films by hybrid sputtering. J. Phys. Chem. Solids **66**(11), 1978–1981 (2005)
9. Mali, S.S., Shinde, P.S., Betty, C.A., Bhosale, P.N., Oh, Y.W., Patil, P.S.: Synthesis and characterization of Cu_2ZnSnS_4 thin films by SILAR method. J. Phys. Chem. Solids **73**(6), 735–740 (2012)
10. Li, J., Chen, G., Xue, C., Jin, X., Liu, W., Zhu, C.: $Cu_2ZnSnS_{4-x}Se_x$ solar cells fabricated with precursor stacked layer ZnS/Cu/SnS by a CBD method. Sol. Energy Mater. Sol. Cells **137**, 131–137 (2015)
11. Kumar, Y.K., Babu, G.S., Bhaskar, P.U., Raja, V.S.: Preparation and characterization of spray-deposited Cu_2ZnSnS_4 thin films. Sol. Energy Mater. Sol. Cells **93**(8), 1230–1237 (2009)
12. Pawar, B.S., Pawar, S.M., Shin, S.W., Choi, D.S., Park, C.J., Kolekar, S.S., Kim, J.H.: Effect of complexing agent on the properties of electrochemically deposited Cu_2 $ZnSnS_4$ (CZTS) thin films. Appl. Surf. Sci. **257**(5), 1786–1791 (2010)
13. Li, J., Ma, T., Wei, M., Liu, W., Jiang, G., Zhu, C.: The $Cu_2ZnSnSe_4$ thin films solar cells synthesized by electrodeposition route. Appl. Surf. Sci. **258**(17), 6261–6265 (2012)
14. Paraye, A. Sani, R., Ramachandran, M., Selvam, N.V.: Effect of pH and sulfur precursor concentration on electrochemically deposited CZTS thin films using glycine as the complexing agent. Appl. Surface Sci. **435**, 1249–1256 (2018)
15. Nguyen, D.C., Ito, S., Dung, D.V.A.: Effects of annealing conditions on crystallization of the CZTS absorber and photovoltaic properties of Cu(Zn, Sn) (S, Se)2 solar cells. J. Alloy. Compd. **632**, 676–680 (2015)

Chapter 2
Production of Bioethanol from Banana Peel Using Isolated Cellulase from *Aspergillus Niger*

Indulekha John, Prasanthi Yaragarla and Arunagiri Appusamy

Abstract In the present work, banana peel, being one of the least investigated biomasses in India, was utilized for the production of bioethanol. Dried and ground banana peel was subjected to ultrasonication at an operating frequency of 20 kHz and 750 W with 2% (v/v) sulphuric acid for 1 h. After the pretreatment, the lignin reduction was 67.50% and the hemicellulose and cellulose recoveries were 45.6 and 69.3%, respectively. *Aspergillus niger* which is capable of producing cellulase enzyme was utilized for the hydrolysis of pretreated banana peel. The activity of isolated crude cellulase enzyme was estimated as 1.712 FPU/ml. The pretreated banana peel was hydrolysed at 50 °C, and the effect of enzyme loading [10–50 FPU] and incubation time [24–96 h] on reducing sugar was determined. Hydrolysate was fermented using 24 h activated *Saccharomyces cerevisiae*, and the effect of fermentation time from 18 to 120 h was investigated. Ethanol concentration was maximum at 24th h of fermentation, and it was 4.24 g/l. The high recovery of cellulose via ultrasonication makes the banana peel suitable for bioethanol production. By providing the in vitro produced enzymes, the usage of commercial enzymes can be eliminated and hence the cost of production of bioethanol might be reduced.

Keywords Banana peel · Bioethanol · Cellulase · Fermentation · Hydrolysis

2.1 Introduction

The world has been depending on fossil fuels as a major energy source. Despite the fact that fossil fuel overwhelms the energy generation, renewable energy resource like biomass is achieving recognition because of the execution of energy policies and better reception on the significance of green energy [1]. Bioethanol as a fuel has become good alternative to gasoline. Among the liquid biofuels, bioethanol is the most commonly used one [2]. Bioethanol reduces the environmental pollution as

I. John · P. Yaragarla · A. Appusamy (✉)
Department of Chemical Engineering, National Institute of Technology Tiruchirappalli, Tiruchirappalli 620015, Tamil Nadu, India
e-mail: aagiri@nitt.edu

© Springer Nature Singapore Pte Ltd. 2020

V. Sivasubramanian and S. Subramanian (eds.), *Global Challenges in Energy and Environment*, Lecture Notes on Multidisciplinary Industrial Engineering, https://doi.org/10.1007/978-981-13-9213-9_2

well as the consumption of crude oil. There is no net increase in carbon dioxide in the atmosphere when bioethanol is burned [3]. For bioethanol production at industrial scale, it is necessary to choose inexpensive and easily available substrates for fermentation [4]. Bioethanol can be produced from agricultural feedstock, forest residues and dedicated crops. Nevertheless, different bioresources should be explored to evade the over-dependency on a single bioresource for energy generation. The conversion of lignocellulosic biomass to bioethanol is a promising methodology due to its low cost, economic benefit, unconsumed biomass, and it does not disturb the land used for agriculture [5]. Lignocellulose consists of polysaccharides such as hemicellulose, cellulose and also aromatic polymer lignin. A variety of lignocellulosic biomass like rice husk, sugar cane bagasse and corn husk are most utilized for second-generation bioethanol production.

Large quantity of unavoidable solid residues are generated during food processing. It is rather difficult to get reliable data on the amount of wastage or by-product fractions. In general, biomass waste from the food processing has been dumped. This is no more a sustainable process since the high amount of carbohydrates, protein, fats and mineral salts present in these residues cause the formation of some other unwanted compounds, foul smell and water browning [6]. The most common solution to their disposal problems is to use them as such for animal feed or as fertilizers. New methods have been introduced in the recovery, bioconversion and utilization of the valuable constituent present in the food processing industries. In recent years, efforts have been taken towards the utilization of cheap renewable agricultural resources as alternative for ethanol production.

Banana is the major foodstuff in the world after rice, corn and milk [7]. India is the world's leading producer nation for bananas, and it records 27% of the overall production worldwide [8]. 30 - 40% of the total weight of banana is wasted as banana peel. In spite of the fact that banana peel is a fruit waste material, it comprises proteins, fibres and carbohydrates in major quantities. The peels of banana are often dumped in landfills, rivers and unregulated dumping grounds. Therefore, the sustainable utilization of banana peel waste would help to diminish the pollution problems caused by their disposal. Production of bioethanol by using banana waste can be an effective utilization of residual biomass mainly in southern part of India. Banana peels are readily available as agricultural waste that can be exploited for bioethanol production since it is rich in carbohydrates.

Bioethanol production from lignocellulose comprises pretreatment, hydrolysis and fermentation processes. Cell wall of lignocellulose is generally disrupted by the pretreatment process, and thus the polysaccharides become accessible for hydrolysis. Polysaccharides such as hemicellulose and cellulose are converted into monosaccharides in the hydrolysis process. In the fermentation process, these monosaccharides are converted to bioethanol with the help of micro-organisms. The selection of appropriate pretreatment method to remove lignin from cell wall and release more polysaccharides without any loss is a challenging step in bioethanol production process [9]. Also, the high cost of commercial enzymes used in the hydrolysis will not make the process economically feasible. Bioethanol production process can be made sustainable and profitable by using cheaper raw materials and cost-effective in house enzyme production.

Cellulase is the most important enzyme used in the enzymatic hydrolysis of bioethanol production. Cellulase is a class of enzyme that catalyses the cellulolysis [10]. The type of micro-organism which can produce cellulase enzymes is mostly fungi. Among various species of fungi, *Aspergillus* sp. are well known for their ability to produce enzymes [11]. In the present study, *Aspergillus niger* was experimented for the production of cellulase using banana peel as substrate. *A. niger* is generally regarded as non-pathogenic fungus that is widely distributed in nature and became an industrially used micro-organism. The main aim of the study was to convert the banana peel waste into bioethanol using in vitro produced cellulase so as to reduce the cost of production of bioethanol.

2.2 Experimental

2.2.1 Materials

Banana peel was collected from canteen located at NIT Tiruchirappalli campus. Peel was washed and then sun-dried for 3 days. Dried peel was powdered and particle size in the range of 0.5–1 mm was chosen for further studies.

2.2.2 Production of Cellulase

2.2.2.1 Inoculum Preparation of *A. niger*

A. niger was purchased from Microbial Type Culture Collection (MTCC), Chandigarh. Czapek yeast extract (CYA) broth medium was used for the inoculation of *A. niger*. Czapek concentrate was made up of $NaNO_3$: 30 g, KCl: 5 g, $MgSO_4.7H_2O$: 5 g, $FeSO_4$. $7H_2O$: 0.1 g and 100 ml of distilled water. CYA medium was comprised of yeast extract: 5 g/L, K_2HPO_4: 1 g/L, sucrose: 30 g/L and Czapek concentrate: 10 ml/L. Few drops of sterile distilled water was provided to the freeze-dried culture of *A. niger*, and this suspension was streaked on CYA broth medium. *A. niger* was subcultured onto CYA slants, and the slants were incubated for 7 days at 30 °C.

2.2.2.2 Cellulase Production by Solid-State Fermentation

The mineral salt medium used for cellulase production had a composition in g/100 g of substrate: $CoCl_2$, 0.1; $CuSO_4.5H_2O$, 0.5; KH_2PO_4, 5; corn steep liquor, 50; and yeast extract, 0.5. A pH of 7.0 was preferred for the growth of micro-organism.

Banana peel was used as substrate for cellulase enzyme under solid-state fermentation. To a 250-ml Erlenmeyer flask, 30 ml of mineral salt medium and 10 g of substrate were added. The flask was sterilized for 15 min at 121 °C. After sterilization, the flask was cooled down to room temperature and then inoculated with 5 ml of culture. The contents were mixed well and incubated at 30 °C under static conditions for 5 days. The flask was mixed periodically by gentle shaking.

At the end of solid-state fermentation, the enzyme was extracted from the banana peel by mixing homogenously with 50 ml of distilled water. The extract was kept in an orbital shaker rotating at speed of 120 rpm for 1 h. The extract was centrifuged at 6000 rpm for 20 min at 4 °C to remove the debris and the supernatant was used as extracellular cellulase enzyme. The activity of cellulase was assayed according to filter paper assay. Cellulase activity was determined as filter paper hydrolysing activity using a Whatman No.1 filter paper strip of 1 × 6 cm. Enzyme sample was supplemented with 1 ml of 50 mM of sodium citrate buffer of pH 4.8, 50 mg of filter paper strip was immersed in the mixture and then the mixture was incubated at 50 °C for 1 h. After that, 3 ml of dinitrosalicylic acid (DNS) solution was added to the enzyme sample, blank and enzyme blank, and boiled for 5 min; 20 ml of distilled water was added to all the test tubes and the mixture was allowed to settle down by keeping it at room temperature for 20 min. Supernatant liquid was measured against 540 nm for estimation of glucose concentration.

$$\text{FPU} = \frac{0.37}{\text{Enzyme concentration to release 2 mg of glucose}}$$

2.2.3 *Production of Bioethanol from Banana Peel*

2.2.3.1 Pretreatment of Banana Peel by Acid-Assisted Ultrasonication

Powdered banana peel was pretreated by acid-assisted ultrasonication. Ultrasonication experiments were conducted in a jacketed glass vessel. Cold water is distributed through the jacket of the vessel to endure the temperature of the process. Ultrasonic processor used in the present study was probe-type processor (VCX 750, Sonics and Material Inc, USA) which operates at 750 W and 20 kHz. The powdered banana peel was sonicated with 2% (v/v) of sulphuric acid at a solid-to-liquid ratio of 1:10 for 60 min. After cooling down the contents to room temperature, it was filtered. The solid contents attained after filtration were washed repeatedly with distilled water to get a neutral pH for the contents. It was then dried at 45 °C. The composition of the untreated and pretreated banana peel was evaluated to know the effectiveness of pretreatment. The cellulose and hemicellulose contents were estimated using Van Soest fibre analysis, and lignin content was determined using NREL method.

2.2.3.2 Enzymatic Hydrolysis

Pretreated banana peel was hydrolysed by enzymatic method using isolated cellulase enzyme from *A. niger*. To a 250-ml Erlenmeyer flask, the pretreated peel and sodium citrate buffer were added and sterilized at 121 °C for 15 min. After sterilization, the medium was cooled to room temperature and the isolated enzyme was added. The enzyme concentration was varied in each flask, and hydrolysis was carried at 50 °C and in an orbital shaker rotating at speed of 120 rpm. The effect of time on hydrolysis was also investigated for a time period of 24–96 h. The samples were drawn for analysing the reducing sugar content at regular time intervals. The amount of reducing sugar in the samples was estimated by DNS method.

Freshly prepared DNS reagent (3 ml) was added to 3 ml of sample. The mixture was heated at 90 °C for 10 min to develop a red–brown colour; 1 ml of 40% of Rochelle salt was added to the sample for the stabilization of the colour. After cooling down, the sample was analysed at 575 nm in UV–Vis spectrophotometer (Schimadzu UV2600). The amount of reducing sugar in the sample was calculated from the standard curve.

2.2.3.3 Fermentation

Baker's yeast grains (1 g) were inoculated into a 20 ml of sterile sucrose medium in order to activate *Saccharomyces cerevisiae*. It was incubated at 30 °C in an orbital shaker at 100 rpm for 20 h. After activation, the cells were grown on YEPD medium composed of 10 g/L yeast extract, 20 g/L peptone, 20 g/L dextrose and 20 g/L agar at 30 °C for 7 days. Spores were allowed to grow in agar slants for the source of inoculum for bioethanol production.

The hydrolysate obtained after enzymatic hydrolysis was sterilized for fermentation. After cooling down, the medium was inoculated with 5% (v/v) of *S. cerevisiae*. Fermentation of the enzymatic hydrolysate of pretreated banana peel was experimented at 37 °C with an agitation of 100 rpm in an orbital shaking incubator. Samples were drawn at regular time intervals and centrifuged at 6000 rpm for 15 min. The supernatant was analysed for ethanol content by gas chromatography.

2.3 Results and Discussion

2.3.1 Cellulase Production for Bioethanol Production

Cellulase was produced in vitro by solid-state fermentation using *A. niger* with the intention of minimizing the bioethanol production cost by eliminating the commercial cellulase. Since banana peel was chosen as the substrate for bioethanol production, the same was opted as the source of carbon for the production of cellulase. Hence, agro-waste materials such as lignocellulosic biomass can be explored for the production of bioethanol as well as enzymes which has to be utilized in enzymatic hydrolysis of bioethanol production. The cellulase activity of the crude enzyme produced in the present study was determined by filter paper assay and expressed as filter paper units, FPU. One filter paper unit of cellulase is the amount of enzyme that forms 2 mg of glucose for 1 min during hydrolysis reaction. The concentration of enzyme which would release exactly 2.0 mg of glucose was estimated by plotting glucose liberated against enzyme concentration on a semi-logarithmic graph paper. The extracellular crude cellulase activity was observed as 1.712 FPU/ml.

2.3.2 Effect of Pretreatment on the Composition of Banana Peel

Acid-assisted ultrasonication was performed as a pretreatment process for bioethanol production using banana peel. For the production of bioethanol from lignocellulosic biomass, biomass with minimal lignin content and high cellulose content is preferable. The raw banana peel composition is given in Table 2.1. Banana peel has noticeable amount of polysaccharides which include cellulose and hemicellulose. The key factor in using this banana peel is that it has low amount of lignin when compared to other lignocellulosic biomass.

The raw banana peel was sonicated for 60 min with 2% (v/v) sulphuric acid. The composition of pretreated banana peel is also given in Table 2.1. After the pretreatment, there was an adequate increase in the cellulose and hemicellulose content of the banana peel. Moreover, there was a reduction of lignin which is known as delignification of 67.5% in the pretreated peel; 45.6% of hemicellulose and 69.3% of cellulose were recovered in the pretreated peel. When the lignin content is low, cellulose can be easily attacked by cellulase in the enzymatic hydrolysis resulting in the release of glucose which is converted into bioethanol by micro-organisms. Hence, pretreatment plays an important role in bioethanol production. In this aspect, acid-assisted ultrasonication is an effective pretreatment method for banana peel.

The main advantage of dilute acid pretreatment related to other methods is the high recovery of sugars from hemicellulose and it is not only solubilizing the hemicelluloses but also converting the hemicelluloses into fermentable sugars [12]. Sonication can help in the disruption of the bonds, which leads to the formation of macro-radicals and these macro-radicals along with $\dot{H}$ and $O\dot{H}$ radicals stimulate the depolymerization of lignocellulosic material [13]. Hence, the combination of ultrasonication

Table 2.1 Composition of banana peel before and after pretreatment

Components (%)	Before pretreatment	After pretreatment
Cellulose	15.14	25.63
Hemicellulose	11.32	16.48
Lignin	6.40	2.08

and dilute acid pretreatment of banana peel resulted in depolymerization of lignin and recovery of cellulose and hemicellulose.

2.3.3 *Enzymatic Hydrolysis*

Enzymatic hydrolysis was performed on acid-assisted ultrasonication pretreated banana peel using crude cellulase enzyme produced by *A. niger* in the present study. Hydrolysis studies were conducted at 50 °C and 150 rpm, and enzyme loading from 10 to 50 FPU and hydrolysis time from 24 to 96 h were varied. The effect of enzyme loading and hydrolysis time on reducing sugar obtained after hydrolysis was investigated.

Reducing sugar concentration was observed increasing with hydrolysis time from 24 to 96 h as shown in Fig. 2.1. Though maximum reducing sugar was obtained at 96th h, there was not much difference in the reducing sugar content between the hydrolysis times of 72 and 96 h. Figure 2.1 also depicts the effect of enzyme loading on reducing sugar, and it was noted that there was no appreciable amount of reducing sugar even at higher hydrolysis time. This is due to the insufficient amount of enzyme to act upon substrate for hydrolysis. As the enzyme loading increased up to 40 FPU, there was a drastic improvement in the reducing sugar concentration which in turn shows the efficient enzymatic hydrolysis of pretreated banana peel by isolated cellulase. However, reducing sugar was found to be decreasing for an enzyme loading of 50 FPU despite the hydrolysis time. This could be due to the product inhibition mainly by glucose [14]. Hence, the maximum reducing sugar was

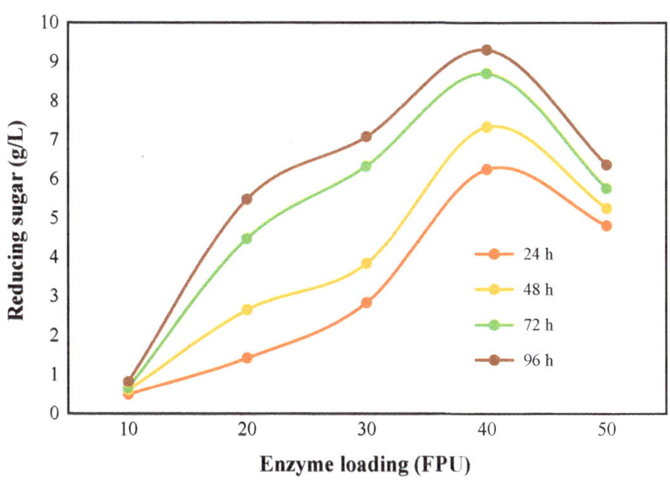

Fig. 2.1 Effect of hydrolysis time and enzyme loading on reducing sugar

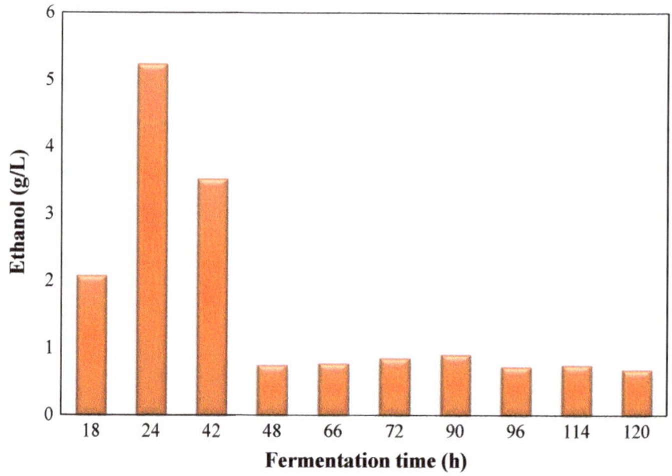

Fig. 2.2 Effect of fermentation time on ethanol

attained after the enzymatic hydrolysis of pretreated banana peel for 96 h using 40 FPU of isolated cellulase and it was 9.3 g/L.

2.3.4 Fermentation

The hydrolysate of pretreated banana peel containing 9.3 g/L of reducing sugar underwent fermentation using *S. cerevisiae*, and the effect of fermentation time on bioethanol was studied. Figure 2.2 shows the effect of fermentation time on bioethanol production. Bioethanol concentration was increased up to 24 h of fermentation time and there was a decrease in the production of bioethanol after 24 h. Experimentally, the maximum bioethanol produced in the present study is 4.24 g/L at 24 h of fermentation. Theoretically from 9.3 g/L of reducing sugar, 4.75 g/L of bioethanol can be produced. Thus, the yield of bioethanol production was recorded as 89% when compared to the theoretical yield. Yield coefficient of bioethanol production in terms of reducing sugar consumption was 0.455.

2.4 Conclusion

Banana peel was experimented for the production of bioethanol. The combination of dilute acid hydrolysis and ultrasonication pretreatment was very effective to release more cellulose and hemicellulose and also to remove lignin which is mandatory for bioethanol production. With 2% sulphuric acid and ultrasonication for 1 h, 69.3%

cellulose was recovered and 67.5% lignin was removed. Crude cellulase enzyme was produced from *A. niger* using banana peel as substrate. The activity of cellulase was determined as 1.712 FPU/ml. Isolated cellulase was utilized in the hydrolysis of pretreated banana peel. Reducing sugar was increased with hydrolysis time and enzyme loading up to 40 FPU. Further increase in enzyme loading decreased the reducing sugar concentration. Fermentation was performed using *S. cerevisiae*. The maximum ethanol was obtained at 24 h of fermentation, and it was 4.24 g/L corresponding to a yield of 89%. Thus, acid-assisted pretreatment and enzymatic hydrolysis by isolated cellulase could result in good amount of bioethanol. Bioethanol production from banana peel can be more explored and enzyme production can be optimized for large-scale productions.

Acknowledgements The authors of this manuscript wish to thank National Institute of Technology Tiruchirappalli for providing a platform to carry out the research work and the Ministry of Human Resource Development for granting doctoral research fellowship and GATE fellowship to Ms. Indulekha John and Ms. Yaragarla Prasanthi.

References

1. Tock, J.Y., Lai, C.L., Lee, K.T., Tan, K.T., Bhatia, S.: Banana biomass as potential renewable energy resource: a Malaysian case study. Renew. Sustain. Energy Rev. **14**(2), 798–805 (2010)
2. John, I., Siddarth, M.S.G., Kalaichelvi, P., Arunagiri, A.: Characterization of citrus peels for bioethanol production. In: Raj Mohan, B., Srinikethan, G., Meikap, B.C. (eds.) Materials, Energy and Environment Engineering, pp. 3–12. Springer (2017)
3. Demirbas, A.: Bioethanol from cellulosic materials: a renewable motor fuel from biomass. Energy Sources **27**(4), 327–337 (2005)
4. Cardona, C.A., Quintero, J.A., Paz, I.C.: Production of bioethanol from sugarcane bagasse: status and perspectives. Biores. Technol. **101**(13), 4754–4766 (2010)
5. John, I., Muthukumar, K., Arunagiri, A.: A review on the potential of citrus waste for D-Limonene, pectin, and bioethanol production production. Int. J. Green Energy **14**(7), 599–612 (2017)
6. Marriott, N., Gravani, R.B.: Principles of Food Sanitation. Springer Science & Business Media (2006)
7. Sharma, P., Mishra, A.A.: Biofuel production from banana peel by using micro wave. Int. J. Sci. Eng. Technol. **3**(4), 1015–1018 (2015)
8. Kishor Kumar, M., Muralidhara, B.M. Usha Rani, M., Akshita Gowda, J.: A figuration of banana production in India. Environ. Ecol. **31**(4A), 1860–1862 (2013)
9. John, I., Yaragarla, P., Muthaiah, P., Ponnusamy, K., Appusamy, A.: Statistical optimization of acid catalyzed steam pretreatment of citrus peel waste for bioethanol production. Res. Efficient Technol. **3**, 429–433 (2017)
10. Mahalakshmi, N., Jayalakshmi, S.: Cellulase production by *Aspergillus niger* under solid state fermentataion using agro industrial wastes. Int. J. Adv. Multi. Res. **3**(2), 78–83 (2016)
11. Sridevi, A., Narasimha, G., Reddy, B.R.: Production of cellulase by *Aspergillus niger* on natural and pretreated lignocellulosic wastes. Internet J. Microbiol. **7**(1), 580–592 (2009)
12. Arumugam, R., Manikandan, M.: Fermentation of pretreated hydrolyzates of banana and mango fruit wastes for ethanol production. Asian J. Exp. Biological Sci. **2**(2), 246–256 (2011)
13. Rehman, M.S.U., Kim, I., Chisti, Y., Han, J.: Use of ultrasound in the production of bioethanol from lignocellulosic biomass. Energy Educ. Sci. Technol. Part A Energy Sci. Res. **30**(2), 1391–1410 (2013)

14. Ballesteros, I., Negro, M.J., Oliva, J.M., Cabañas, A., Manzanares, P., Ballesteros, M.: Ethanol production from steam-explosion pretreated wheat straw. Appl. Biochem. Bitechnology **129**(132), 496–508 (2006)

Chapter 3
Performance Evaluation of Two-Stage Vertical Flow Multispecies Constructed Wetland for Domestic Wastewater Treatment

G. R. Munavalli, P. G. Sonavane, G. V. Deshpande, S. M. Wandare and N. M. Aswale

Abstract Decentralized treatment of domestic wastewater with the use of constructed wetland (CW) as physical and biological treatment is widely used. The performance of CW can be enhanced by using multispecies and dual support medium. In this context, a laboratory-scale treatment system consisting of a settling tank followed by two-stage vertical flow multispecies constructed wetland (VFMCW) was developed. The reactors of VFMCW connected in series were planted with *Canna India* and *Phragmites Karka* in charcoal and grit as support medium. The system was operated in continuous fill and drain mode for various hydraulic loading rate (HLR) which corresponds to hydraulic retention time (HRT) of 1.50–12 h for each stage and overall system. The feed used for system was wastewater collected from sump which is contributed by greywater and blackwater from hostel in the college (Walchand College of Engineering, Sangli) premises. The performance of the system was evaluated for organic matter (measured in terms COD) removal. Dissolved oxygen (DO) enhancement in the system was also assessed. The COD removal efficiency was found to be 53–80% for the HLR varied from 60 to 7.5 mm/h. Both the stages of VFMCW contributed significantly to DO in effluent for all HLRs.

3.1 Introduction

The consumption of water and generation of domestic wastewater are increasing with time. The existing collection and treatment facilities are inadequate to satisfy the desired requirements of safe disposal. There is a large gap between domestic wastewater generated and treated. National θSample Survey (NSS) shows that in 2012 nearly 49.9% of rural households and 12.5% of urban households of India had no drainage arrangement [12]. Septic tank and soakage/disposal pits are more common to treat black water in un-sewered localities. Greywater is disposed in open drains

G. R. Munavalli (✉) · P. G. Sonavane · G. V. Deshpande · S. M. Wandare · N. M. Aswale
Department of Civil Engineering, Walchand College of Engineering, 416415 Sangli, Maharashtra, India
e-mail: guru.munavalli@walchandsangli.ac.in

© Springer Nature Singapore Pte Ltd. 2020

V. Sivasubramanian and S. Subramanian (eds.), *Global Challenges in Energy and Environment*, Lecture Notes on Multidisciplinary Industrial Engineering, https://doi.org/10.1007/978-981-13-9213-9_3

and finally makes its way to natural nallas. The treated wastewater is not disposed of properly in the areas where proper collection and treatment system exist. This has caused a lot of pollution problems on the sources of surface water and groundwater, degradation of aquatic ecosystem, and impairment of water quality for drinking. Domestic wastewater can be collected from the entire area to a common point and treated centrally. These systems require huge infrastructure, investment, and skill in operation. The complexities increase with size of the system and require mechanized systems to address these complexities. Decentralized wastewater treatment system (DWTS) can be another option suitable to address these issues and provide an opportunity for public to share the responsibility of waste management. These systems are cost-effective and less energy intensive, and concepts of natural purification can be induced in the treatment system. DWTS is an onsite system that is used to treat relatively small volumes of wastewater, normally generated from individual or groups of households. The appropriate technologies for DWTS include modified septic tank, anaerobic filter, anaerobic baffled reactor, moving bed bioreactor, vermin filters, and constructed wetland. DWTS based on combination of anaerobic and aerobic treatment systems is more potential option in Indian context. Constructed wetland (CW)-based DWTS are natural, affordable, eco-friendly, and sustainable for treating domestic wastewater. CWs have been implemented in many configurations, viz. horizontal flow subsurface, vertical flow, free surface, and hybrid. Vertical flow multispecies constructed wetlands (VFMCW) are normally preferred due to their higher oxygen transfer capacity. The studies conducted on VFMCW include [1–9], and these studies have used mono-species and mono-medium substrate in VFMCW systems. In the reference [10], VFCMW system is reviewed in terms of their applicability and treatment potential for removing organic contaminants from wastewater. The studies on configuration modification to reduce the area requirement for VFCMW for domestic wastewater treatment are less. Further, the studies were restricted to single-stage VFMCW system, and the potential of VFMCW operated in two stages is not yet fully explored for organic matter removal. In this context, the present study focuses on the modification of the configuration of VFMCW with multispecies to treat domestic wastewater to enhance organic matter removal. The possible contributions to existing work on VFMCW include use of multispecies and dual media in two-stage vertical flow multispecies constructed wetland (VFMCW).

3.2 Materials and Methods

3.2.1 Source of Wastewater

The wastewater samples were collected from a disposal location on the campus of Walchand College of Engineering, Sangli (Maharashtra). The municipal water supply is used for potable use after retreatment in college campus. The open wells located within the college campus are used to supply water for non-potable use. Thus, the water in bathroom and other sanitary utilities are supplied with groundwater. The wastewater generated from residential bungalows, ladies, and boys' hostel is disposed of through a sewer to a disposal site. The typical usage of water from these buildings includes bathing, washing, and toilet flushing. The hostels are provided with septic tanks for treatment of blackwater and effluent from septic tank flows into sewer. The greywater (from bathroom and washings) flows directly into sewers. Grab sampling procedure was used to collect the samples in morning hours (8–10 am) in order to ensure uniform characteristics.

3.2.2 *Experimental Setup*

A laboratory-scale experimental setup consisting of pretreatment and two-stage VFMCW was developed and is shown in Fig. 3.1.

The two stages are arranged vertically one below the other resulting in cascade-type system. This type of arrangement provides more surface area in a given space. The stage 1 and stage 2 VFMCW reactors are made up of circular plastic tanks having diameters 32 and 60 cm, respectively. The surface areas of the first and second stages are 0.08 and 0.264 m^2, respectively. Settling tank is provided as a pretreatment to remove suspended solids from raw wastewater. The flow is regulated at all the stages (settling tank, VFMCW) through control valves. This arrangement helps in adjusting required HLR. The substrate/support medium used includes grit, lignite coal, and coarse gravel. The coal is used due to its adsorptive nature. The outlet in second-stage VFMCW is located in such a way that there is no short circuit of flow, and the flow is channelized irrespective of its point of entry. The flow distributors made up of perforated flexible pipes were provided to uniformly spread the wastewater in both stages of VFMCW. The size and depth of the medium are given in Table 3.1.

Canna indica and *Phragmites karka* were used as vegetation for both the reactors. The vegetation was chosen considering their local availability, tolerance to organic load, and potential of developing good root matrix together. *Phragmites karka* has fibrous roots system with horizontally crawling rhizomes and *Canna indica* with relatively shallow roots.

3.2.3 *Planning and Operation of System*

Experiments were planned to conduct performance evaluation of two-stage VFMCW to remove chemical oxygen demand (COD) and dissolved oxygen (DO) addition. pH was also monitored. The methods of analysis were referred to [11]. The system was assessed for COD removal when operated in fill-drain-type continuous mode. A typical operation consisted of:

i. The settling tank was filled with 30 L raw wastewater and settled for 2 h.

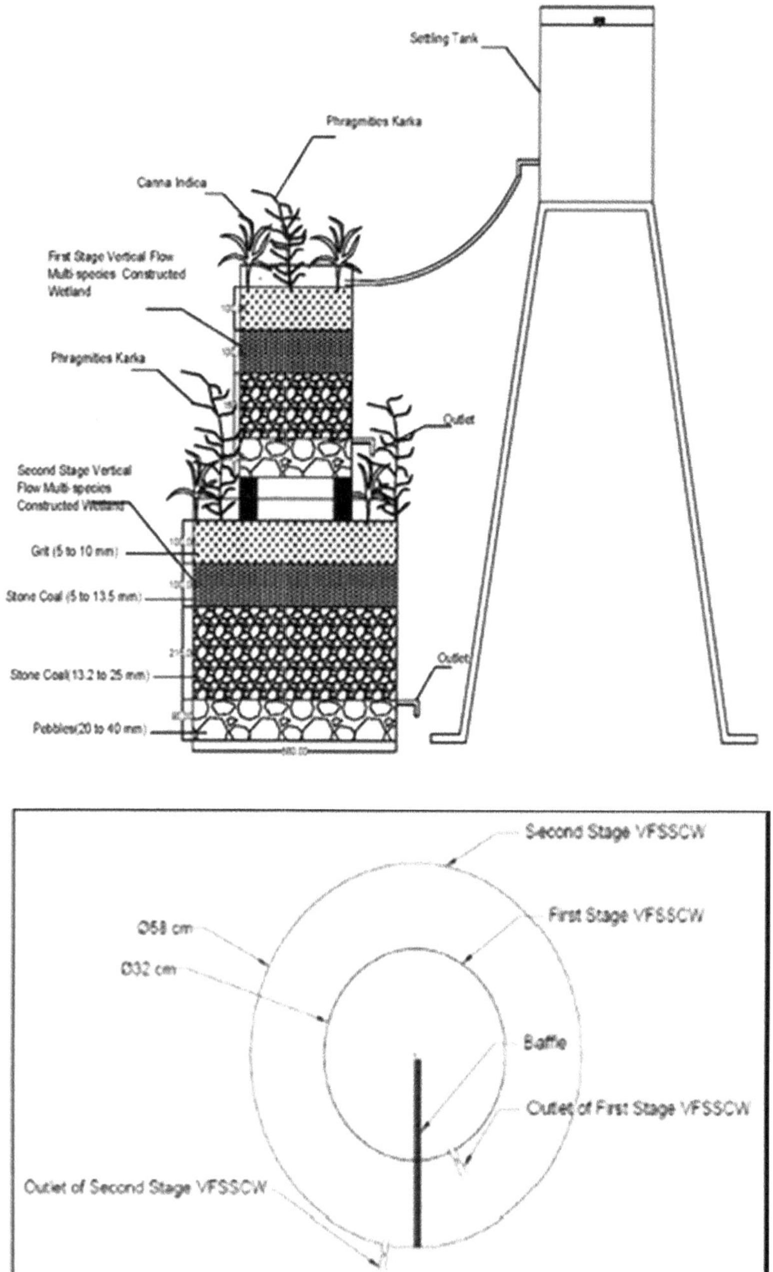

Fig. 3.1 Schematic sketch of experimental setup (VFMCW)

Table 3.1 Details of support medium

Medium	First-stage reactor		Second-stage reactor	
	Size (mm)	Depth (cm)	Size (mm)	Depth (cm)
Grit	5–10	10	5–10	10
Lignite coal	5–13.2	10	5–13.2	10
	13.5–25	10	13.5–25	15
Gravel	20–40	10	20–40	10

Overall porosity 40% for both stages

ii. The flow rate was regulated by adjusting valve at the outlet of respective reactors to achieve the required hydraulic loading rate (HLR) (e.g., 30 L of flow from settling tank distributed onto VFMCW in 1.5 h shall provide an overall HLR of 7.5 mm/h).
iii. The settled wastewater was applied through distributor pipe at this adjusted rate onto VFMCW.
iv. The applied wastewater moves sequentially through stage 1 and stage 2 of VFMCW in simultaneous fill-drain mode continuously.
v. The effluent was collected after settling and both stages of VFMCW for analysis.

The VFMCW was applied with and assessed for HLRs of 7.5, 15, 30, and 60 mm/h. The corresponding organic loading rate (OLR) was 47, 96, 157 and 272 gCOD/m^2·d, respectively, for above HLRs. The respective feed time of 30 L wastewater for these HLRs is 12, 6, 3, and 1.5 h. Further, the variability of effluent quality (COD and DO) for each HLR was also assessed.

3.3 Results and Discussion

3.3.1 Wastewater Characterization

The summary of wastewater characteristics during the period of study is given in Table 3.2. The organic strength of wastewater is low (evident in values of COD and BOD), and hence, it is a medium strength wastewater. The ratio of BOD/COD is more than 0.5 indicating amenability of this wastewater for biological treatment. The higher value of TDS indicates a relatively higher proportion of groundwater being used for non-potable purpose than surface water. TSS and TKN are also observed to

Table 3.2 Wastewater characteristics

Parameter	pH	TDS	TSS	TKN	COD	BOD_5
Value	7.60 ± 0.40	1260 ± 150	175 ± 15	130 ± 15	250 ± 25	135 ± 20

All values are in mg/L except pH

be significant in quantity. The wastewater with these characteristics is not disposable either for land disposal or onto water bodies. The characterization also indicates that settling is required before feeding this wastewater to VFMCW.

3.3.2 *Effect of HLR on Performance of VFMCW*

Figure 3.2 shows the effect of HLR on COD removal. Settling tank contributes by 25–35% for COD removal, and it is suggestive of 2 h settling which is sufficient. The contribution by second-stage VFMCW is slightly higher than that of the first-stage VFMCW for COD removal for all HLRs. The enhancement is found to be about 10% irrespective of change in applied HLR. This is due to larger area, space, and better oxygenation condition available in second-stage VFMCW. The results

Fig. 3.2 Photographic view of experimental setup (VFMCW)

also show that there is a decrease in COD removal when HLR is increased. HLR applied less than 15 mm/h results in higher removal rates. The lower percolation rates are observed when the applied HLR is lower and which provides more HRT for the feed wastewater. It can also be seen that the overall efficiency of the VFMCW system is in the range 60–80% for the applied HLR between 7.5 and 60 mm/h. Figure 3.3 shows the summary of HLR effect and OLR affecting the COD removal. Similar to the observation on HLR, decrease in OLR improves COD removal rate. The applied HLR and OLR are comparable with those reported in the literature. HLR of 15 mm/h and OLR of less than 100 gm COD/m^2·d seem to be appropriate to achieve efficiency around 80%. Majority of COD removal in VFMCW is attributed to aerobic attached growth bacterial action. The surface for bacterial attachment is provided by roots and support medium. The use of multispecies (*Canna indica* and *Phragmites karka*) provided a dense matrix of shallow and deep roots. The oxygen liberated at root surface and simultaneous filling/draining action maintained aerobic conditions within the wetland. This has significantly contributed to higher COD removal. Further, the use of coal as a support medium provided adsorptive surface for COD removal. However, contribution of coal in COD removal was not assessed in this study.

3.3.3 Assessment of Oxygenation in VFMCW

Figure 3.4 shows the oxygenation condition within both the stages of VFMCW. It can be seen from the results that DO levels are in the range 1.5–3.5 mg/L. The addition of DO is observed to be more in second-stage VFMCW than in the first stage. This is due to the fact that volume of second-stage VFMCW is more, and it provides more space for aeration/ventilation within the wetland. Slightly lower values of DO are observed for high HLR runs than that of low HLR. There is an increase in DO level with increase in HLR; however, the increase is not significant. The oxygen absorption is governed by quality of settled wastewater (temperature, impurities, etc.), saturation DO, and time of contact. In the present study, the quality of settled wastewater has governed the oxygen absorption as there was no significant increase in DO for lower HLR (high HRT). Normally, DO value in the range 1.5–2.0 mg/L is desirable for sustaining aerobic bacterial activity. The results obtained in this study indicate that aerobic conditions prevail within the wetland when operated in simultaneous fill and drain mode. As mentioned in the previous section, the source of oxygen addition within VFMCW includes wetland plants releasing oxygen at roots and air movement during fill/drain mode of operation. A contact time of 1.5 h which corresponds to HLR of 60 mm/h was found to be adequate to oxygenate the wetland medium to required minimum DO of 1.50 mg/L (Fig. 3.5).

VFMCW system works normally under unsaturated condition. Frequent and simultaneous feeding and draining maintain proper aerobic conditions within the system. The present study assessed the VFMCW system in series for COD removal and possible DO enhancement naturally. The observed COD removal rates are lower

Fig. 3.3 Effect of HLR on COD removal for various experimental runs

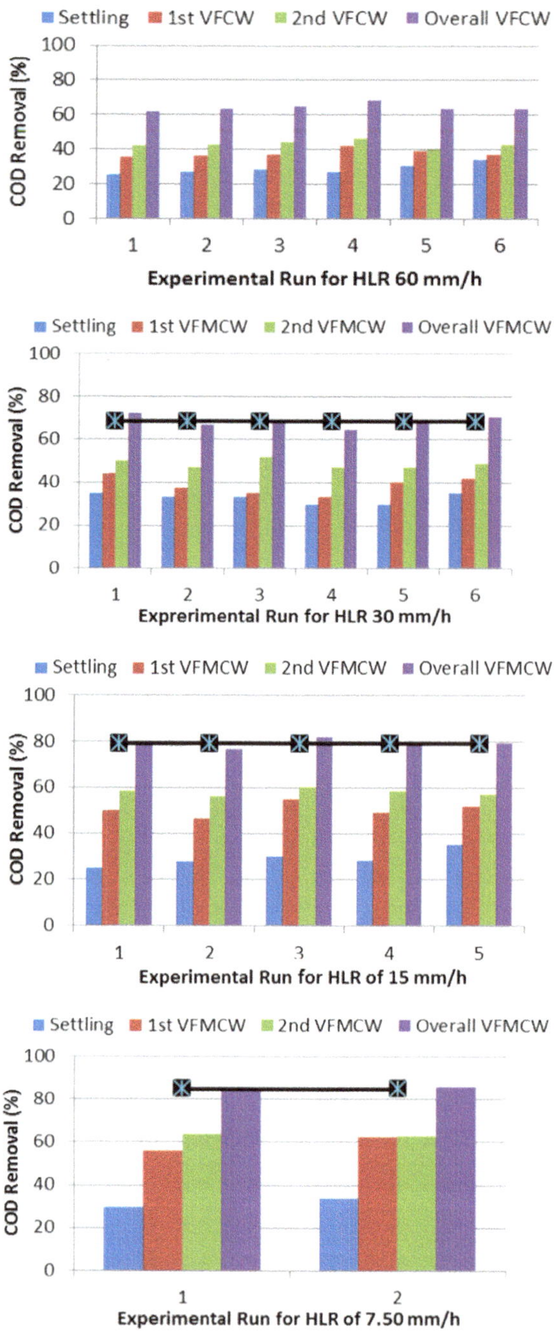

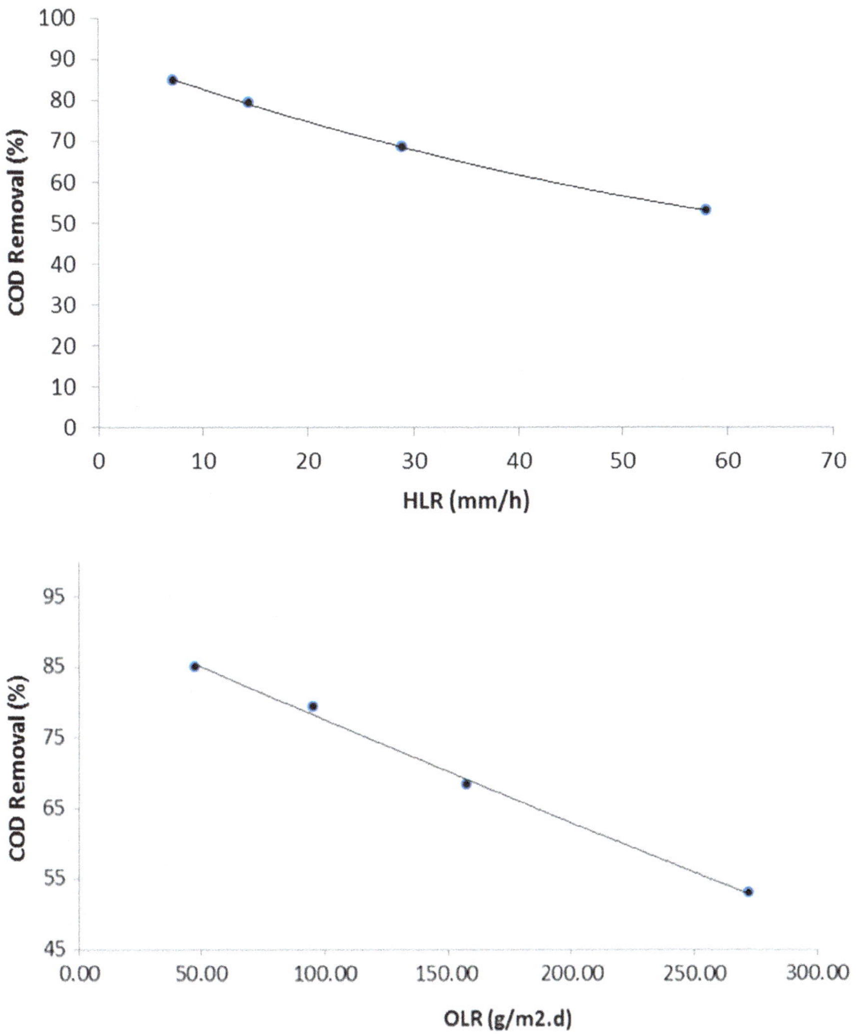

Fig. 3.4 Effect of HLR and OLR on COD removal

in high HLR and OLR runs. Even though proper conditions prevail to support aerobic bacterial activity, the COD removal rate is varying. It suggests that the existence of aerobic activity alone may not be adequate to treat wastewater efficiently. Time of contact and aerobic conditions together govern the COD removal.

Fig. 3.5 Oxygenation in VFMCW

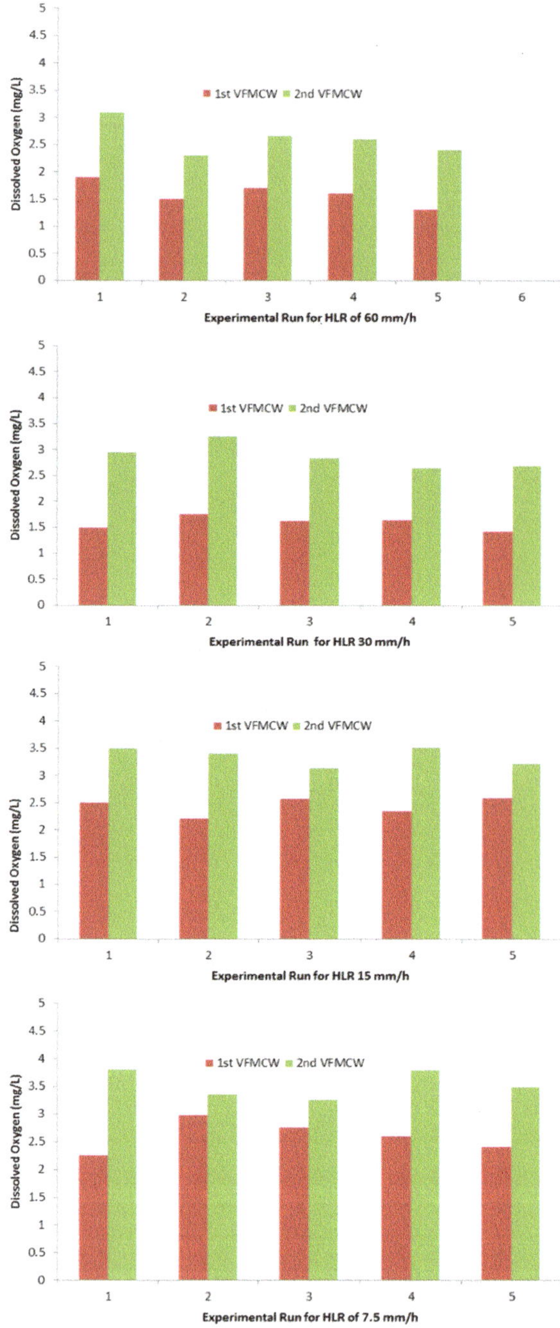

3.4 Conclusions

VFMCW in series was assessed for COD removal and DO enhancement with domestic wastewater as a feed. HLR and OLR were varied to demonstrate the effect of these on COD removal. Based on the study, the conclusions drawn include:

1. Domestic wastewater can be treated by biological treatment.
2. HLR and OLR affect COD removal significantly.
3. Adequate DO conditions prevail within VFMCW to support aerobic activity.
4. Overall COD removal by VFMCW to an extent of 50–85% can be achieved by such systems.

Acknowledgements The present study is a part of research project sanctioned by the Department of Science and Technology (DST) under Water Treatment Initiative (WTI) programme. The authors sincerely acknowledge and thank DST for providing financial support to undertake this study.

References

1. Brix, H.: Arias AC The use of vertical flow constructed wetlands for on-site treatment of domestic wastewater: new Danish guidelines. Ecol. Eng. **25**, 491–500 (2005)
2. Morari, F.: Giardini, L.: Municipal wastewater treatment with vertical flow constructed wetlands for irrigation reuse. Ecol. Eng. **35**, 643–653 (2009)
3. Sklarz, M.Y., Gross, A., Yakirevich, A., Soares, M.I.M.: A recirculating vertical flow constructed wetland for the treatment of domestic wastewater. Desalination 617–624 (2009)
4. Chang, J.-J., Wu, S.-Q., Dai, Y.-R., Liang, W., Wu, Z.-B.: Treatment performance of integrated vertical-flow constructed wetland plots for domestic wastewater. Ecol. Eng. **44**, 152–159 (2012)
5. Abou-Elela, S.I.: Hellal MS Municipal wastewater treatment using vertical flow constructed wetlands planted with *Canna Phragmites* and Cyprus. Ecol Eng **47**, 209–213 (2012)
6. Gikas, G.D.: Tsihrintzis VA A small-size vertical flow constructed wetland for on-site treatment of household wastewater. Ecol. Eng. **44**, 337–343 (2012)
7. Ávila, C., Garfí, M., García, J.: Three-stage hybrid constructed wetland system for wastewater treatment and reuse in warm climate regions. Ecol. Eng. **61**(2013), 43–49 (2013)
8. Ávila, C., Nivala, J., Olsson, L., Kassa, K., Headley, T., Mueller, R.A., Bayona, J.M., García, J.: Emerging organic contaminants in vertical subsurface flow constructed wetlands: influence of media size, loading frequency and use of active aeration. Sci. Total Environ. **494** 211–217 (2014)
9. Tsihrintzis, V.A.: The use of vertical flow constructed wetlands in wastewater treatment. Water Resour Manage **31**, 3245 (2017)
10. Stefanakis, A.I., Akratos, C.S., Tsihrintzis, V.A.: Vertical flow constructed wetlands: eco-engineering systems for wastewater and sludge treatment. Elsevier Science, Amsterdam (2014)
11. APHA.: Standard methods for examination of water and wastewater, 22nd edn. American Public Health Association, Washington, DC (2012)
12. Swachhta Status Report.: National Sample Survey Office, Ministry of Statistics and Programme Implementation, Government of India (2016)

Chapter 4
Comparative and Kinetics Studies of Organo-Nano-Kaolin Clay And Organo-Nano-Bentonite Clay as the Adsorbents for the Reduction of Chromium (VI) from Tannery Effluent

B. Uma Maheswari, V. M. Sivakumar and M. Thirumarimurugan

Abstract Tannery effluents are ranked as the highest pollutants among all industrial wastes. Nowadays, disposal of chromium is one of the major emerging environmental problems in the tanning industry. Cr (VI) is highly toxic and soluble in water, and it is a strong oxidizing agent that causes severe damage to environment. The ability to reduce toxic substances to safe levels effectively and at a reasonable cost is very important. Clay and clay minerals have been used as an effective, low-cost method for the removal of heavy metals from industrial effluents. Organoclays are used as a wide range of adsorbents for removal of heavy metals from wastewater. The main objective of the paper is to examine the feasibility and efficiency of using organo-bentonite nanoparticle (OBN) and organo-kaolin nanoparticle (OKN) as the adsorbents for the removal of Cr (VI) ions from tannery effluent. Adsorption of Cr (VI) ions was carried out on a batch process. Experiments were performed as a function of pH, adsorbent concentration, and contact time. Based on the difference on Cr (VI) ion concentration before and after adsorption, the percentage reduction was calculated. A comparative study is made between OBN and OKN, and maximum reduction efficiency was achieved by OKN particle (95.3%). Isotherm data and kinetic studies were performed. The equilibrium adsorption isotherm was better described by Langmuir adsorption isotherm model. The adsorption kinetics well fitted using pseudo-second-order kinetic model.

Keywords Heavy metal · Organo-bentonite · Organo-kaolin · Isotherm · Kinetics

---

B. Uma Maheswari · V. M. Sivakumar (✉) · M. Thirumarimurugan
Department of Chemical Engineering, Coimbatore Institute of Technology, Coimbatore 641014, India
e-mail: vmsivakumar@gmail.com

© Springer Nature Singapore Pte Ltd. 2020

V. Sivasubramanian and S. Subramanian (eds.), *Global Challenges in Energy and Environment*, Lecture Notes on Multidisciplinary Industrial Engineering, https://doi.org/10.1007/978-981-13-9213-9_4

4.1 Introduction

Due to rapid growth of industries, pollutants such as heavy metals, oxyanions, and organic compounds pollute the environment to a greater extent. Out of all these pollutants, heavy metals are highly toxic and cause adverse effect to the plants, animals, and also human beings. Tannery industry damages the environments to a greater extent by releasing discharge with high oxygen demand, discoloration, and organic and inorganic pollutant. Chromium (VI) is highly toxic and causes series damage to the human life, plants, and animals. In order to treat these heavy metals, various physical, chemical, and biological methods have been studied [1]. Although variety of methods available adsorption is considered as a low-cost and effective method for the treatments of heavy metals [2–4]. A variety of adsorbents such as starch, fly ash, coconut shell, chitin, chitosan, clay, zeolites have been reported [5–11].

The selection of suitable adsorbent is very important because some type of adsorbent can be used to adsorb only some specific type of heavy metal. Clay and clay minerals are used as alternative and inexpensive adsorbent [12]. Clay materials are modified to organically modified clay so that they possess high specific surface area and high cation exchange capacity, good chemical and mechanical stability.

The main objective of the work deals with a comparative study for the reduction of Cr (VI) ions from tannery effluent using organically modified kaolin nanoparticle (OKN) and bentonite nanoparticle (OBN). The effect of pH, adsorbent dosage, contact time was optimized for maximum adsorption capacity by batch experiment.

4.2 Materials and Methods

4.2.1 *Materials*

Bentonite and kaolin clay, provided by HiMedia, were used for the preparation of organoclay. Cetyltrimethylammonium bromide (CTAB) and potassium chromate purchased for Novatech chemical company, India, were used as a surfactant and metal ion source to prepare synthetic effluent.

4.2.2 *Preparation of Organoclay Nanoparticles (OBN, OKN)*

The OBN and OKN were prepared by cation exchange reaction, adapted from the reported procedure [13]. The procedure involves firstly the dispersion of the clay (bentonite and kaolin) in distilled water with agitation (4% w/w) for 30 min; it is then allowed to settle for about 24 h in room temperature. CTAB was added to the suspension, and then, the mixture is stirred for further 20 min and was allowed to settle for 24 h. The mixture was washed thoroughly to remove chloride, till it becomes free from chloride which was confirmed by checking against $AgNO_3$ solution. The final organoclay was dried to 70 °C, sieved at 230 mesh number, and finally ground using a pestle mortar in order to obtain fine powder. The resultant product was characterized by XRD and SEM.

4.2.3 Adsorption Experiments

A stock solution of chromium ions with an initial concentration 400 mg/L was prepared by dissolving fixed amount of potassium chromate into 1 L distilled water. The adsorption experiments were carried out by adding various mass of the adsorbents from 0.5 to 4 grams to a series of conical flask containing 50 mL of metal solution. Subsequently, the flasks were placed in a water bath shaker at 200 r min^{-1} for 1–2 h to reach equilibrium [14]. The metal solution was then centrifuged at 2500 r·min^{-1} for 10 min to remove the solid particles. The residual concentration of Cr (VI) ions in the metal solution was determined by flame atomic absorption (FAA) spectrophotometer at $\lambda_{max} = 540$ nm.

In order to perform kinetic studies, 4 g of OBN and OKN was added to a series of conical flask containing 50 mL of metal solution with initial concentration 400 mg/L. The shaker was then set to 200 r·min^{-1} for a given period of time. At regular time intervals, 10 mL of metal solution was removed from the flask, centrifuged, and analyzed for its residual metal ion concentration. The amount of Cr (VI) ions adsorbed on the solid surface at equilibrium (q_e), in units of mg/g can be determined from the following mass balance equation

$$q_e = \frac{(C_o - C_e)V}{m} \tag{4.1}$$

where C_o and C_e (mg/L) are the concentration of Cr (VI) ions at initial and equilibrium conditions, respectively, V is the volume of the solution (L), and m is the mass of the OBN and OKN (g).

4.2.4 Characterization of Instruments

The X-ray diffraction (XRD) studies of clay and modified clay were performed with XPERT Philips X-ray diffractometer. All the parameters were obtained using Cu/Kα operated at 45 kV, 30 mA. Morphological studies of kaolin, bentonite, OBN, OKN were carried out by scanning electron microscope (SEM) (Philips-EEI ESEM-TEP, Holland). In order to detect the evidence for purity and distribution of specific elements in a solid sample, SEM image was further supported by energy dispersive X-ray (EDX) microanalysis.

4.3 Results and Discussion

4.3.1 Characterization of Adsorbents

XRD studies were carried out in order to identify structural characteristics of the prepared clay and its modified form. The XRD results were reported in Fig. 4.1 representing clay and organically modified clay using CTAB. It was evident from Fig. 4.1 that the spacing expansion and formation of new peak in organically modified clay is due to intercalation of CTAB into clay interlayer. The formation of new peaks in organically modified clay was marked as 1, 2, 3 in Fig. 4.1. Hence, apparently there will be a decrease in hydration water content, which in turn changes the surface

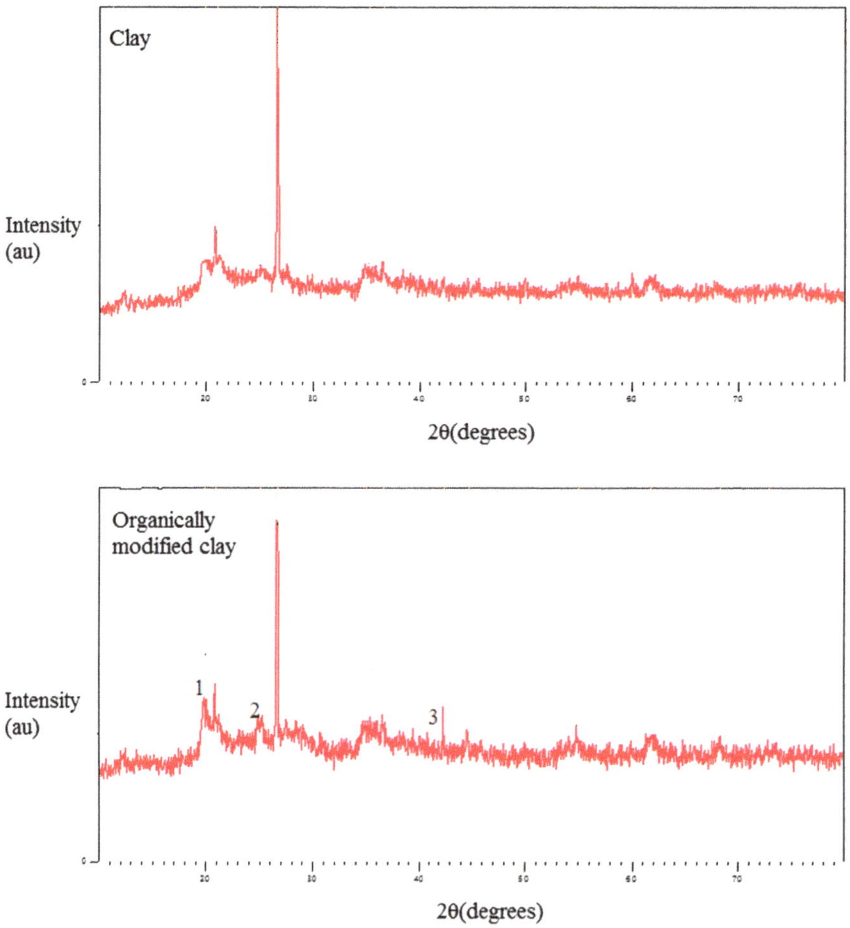

Fig. 4.1 XRD image of clay and organically modified clay

property. A change in the surface property from hydrophilic to hydrophobic was confirmed by shift in peak position.

Similarly, SEM analysis is carried out in order to evaluate the surface morphology of the clay and organically modified clay. The results were depicted in Fig. 4.2a and b. It was clear from Fig. 4.2a that the particles in the clay were closely attached. It

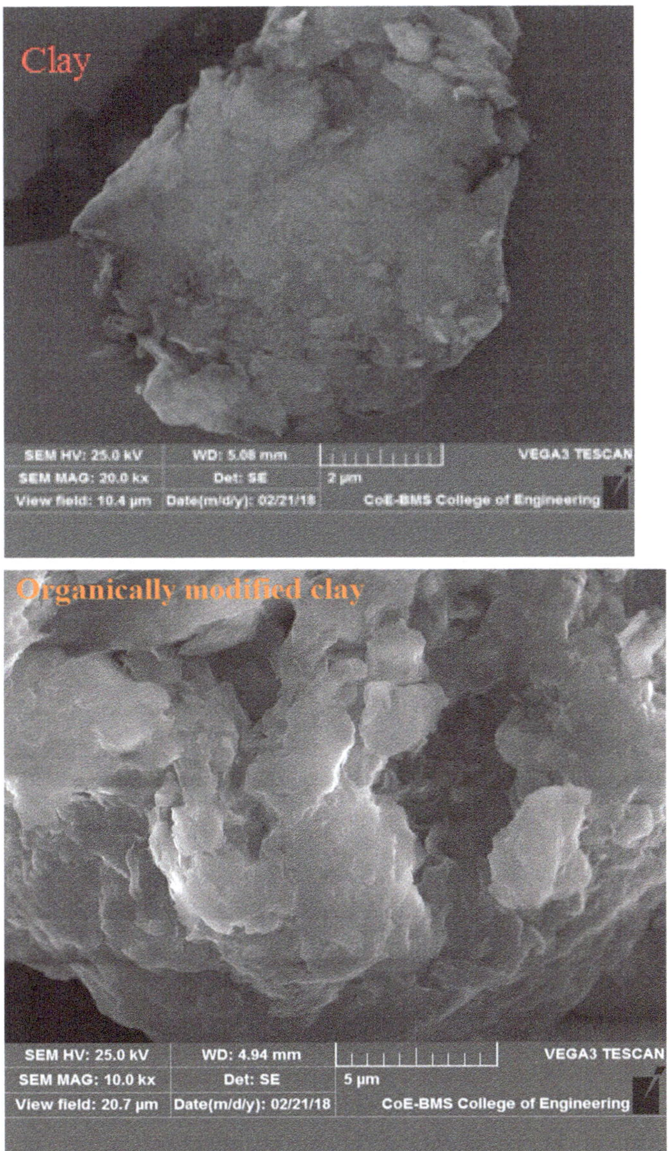

Fig. 4.2 **a** SEM image of clay **b** SEM image of organically modified clay

aggregates due to intermolecular forces. Similarly, from the Fig. 4.2b it is witnessed that CTAB was successfully incorporated in the modified organo clay particles. Dispersion of particles can be clearly observed from Fig. 4.2b compared to Fig. 4.2a.

4.3.2 Effect of pH

Functional group charge of adsorbent, transition metal speciation, and solubility is generally affected by pH. In this study, adsorption experiments were carried out at pH 1–6 range to determine the optimum pH for maximum reduction of chromium (VI) ions. It is observed that increase in pH from 1 to 2, sorption increases slightly for OKN, and maximum Cr (VI) ion adsorption (95.83%) by OKN was observed at pH = 2. Similarly in case of OBN maximum Cr (VI), adsorption (86.1%) was observed at pH = 3. The plot of percentage reduction versus pH for OBN and OKN is shown in Fig. 4.3. The percentage reduction gradually decreases on increase in pH level which can be clearly seen from Fig. 4.3. From the above-obtained results, it clearly indicates that greater number of protons can easily coordinate with the functional groups present on the adsorbents on lower pH [15].

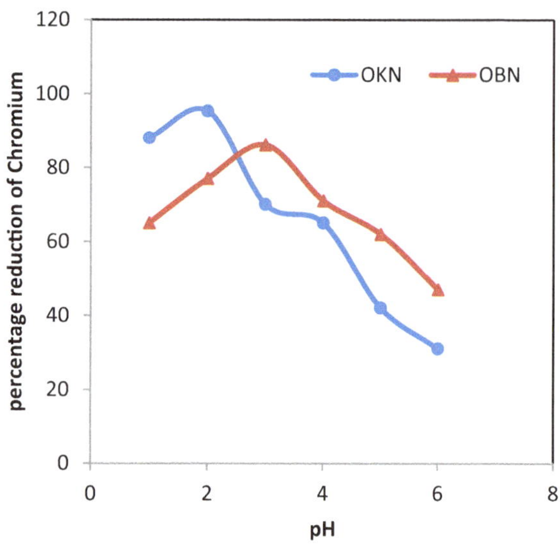

Fig. 4.3 Effect of pH versus percentage reduction of chromium

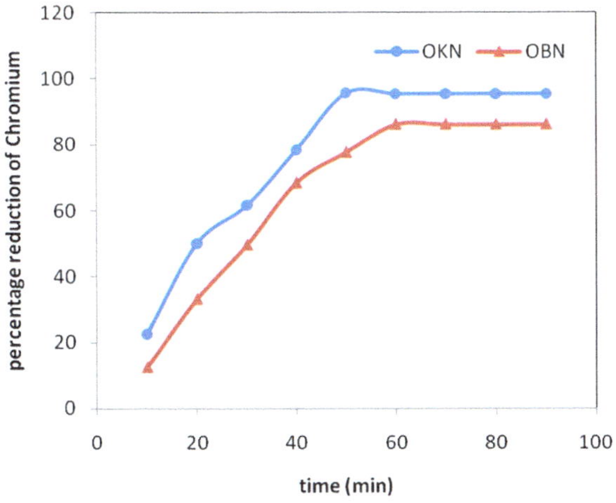

Fig. 4.4 Effect of contact time versus percentage reduction of chromium

4.3.3 *Effect of Contact Time*

The experimental setup measures the effect of contact time on batch adsorption of Cr (VI) ion with an initial concentration of 400 mg/L with a contact time of 20–100 min at a temperature of 25 °C. The reduction of Cr (VI) increased with increase in contact time until the equilibrium point for OBN and OKN. Maximum removal of chromium ions was found at an optimal contact time 60 min for OBN and 50 min for OKN. From the results, it was reported that OKN has very quick uptake of Cr (VI) ions in a short period, by the effect of a large number of vacant sites available on OKN. The results indicate that OKN has high affinity for Cr (VI) ions. The plot of percentage reduction versus contact time is shown in Fig. 4.4.

4.3.4 *Effect of Adsorbent Concentration*

In general, adsorbent concentration specifies the capacity of adsorbent to uptake a certain initial concentration of Cr (VI) ions. Batch adsorption experiments were conducted by varying concentration in the range from 0.5 to 4 g/L for both OBN and OKN. The maximum reduction was about 95.3% at the adsorbent concentration of 4 g/L for OKN and similarly 86.5% for OBN. From Fig. 4.5, it was confirmed that increase in adsorbent dosage increases the percentage of metal sorption; moreover, greater number of OKN creates greater surface area; hence, the number of adsorption sites in the matrix of OKN was increased more when compared to OBN. The plot of percentage reduction versus adsorbent dosage is shown in Fig. 4.5.

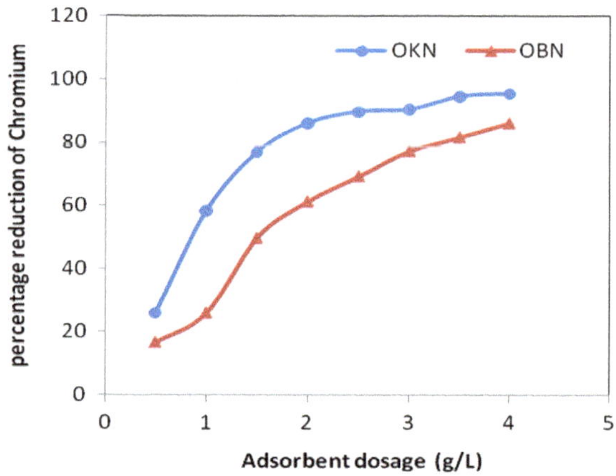

Fig. 4.5 Effect of adsorbent dosage versus percentage reduction of chromium

4.3.5 Equilibrium Studies

In order to optimize the design of a sorption system to Cr (VI) reduction, it is important to establish the most appropriate correlation for the equilibrium curve. Langmuir isotherms have been tested in the present study namely. The goodness of fit between experimental and model predicted values was expressed by the correlation (R^2 values closer or equal to 1).

4.3.5.1 Langmuir Isotherm

Langmuir isotherm has found successful application in many real sorption processes and can be expressed by:

$$q_e = \frac{q_m b C_e}{1 + bC_e} \tag{4.2}$$

The linear form of the Langmuir equation is given by:

$$\frac{C_e}{q_e} = \frac{1}{K_s q_m} + \frac{1}{q_m} C_e \tag{4.3}$$

where C_e is the equilibrium concentration of metal (mg/L), q_e (mg/g) is the Cr^{6+} absorption capacity at equilibrium, q_m (mg/g) is the monolayer adsorbent capacity, and b (L/mg) is the energy constant of adsorption. The plot C_e/q_e against C_e gives straight line shown in Fig. 4.6 which expresses applicability of Langmuir isotherm for OKN and OBN, respectively. The values of q_m and K_s are obtained from slope and intercept of the plot and presented in Table 4.1. The linear representation of langmuir isotherm was shown in Fig. 4.7 and Fig. 4.8 respectively.

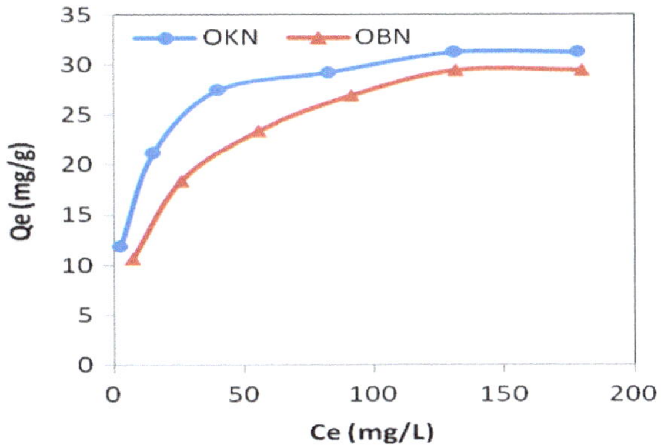

Fig. 4.6 Langmuir isotherm for the reduction of chromium (VI) ion by organo-kaolin nanoparticle (OKN) and organo-bentonite nanoparticle (OBN)

Table 4.1 Langmuir isotherm

Adsorbent	Langmuir isotherm		
	q_m(mg/g)	**b**(L/mg)	R^2
OBN	35.71	0.046	0.993
OKN	26.31	0.145	0.999

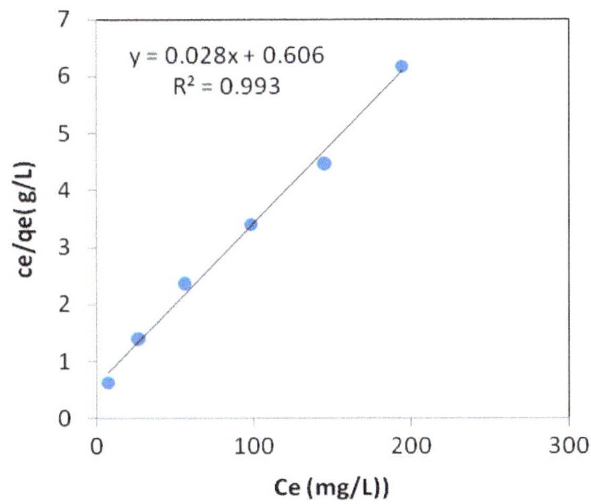

Fig. 4.7 Langmuir isotherm for the reduction of Chromium (VI) ion by organo-bentonite nanoparticle (OBN)

4.3.6 Kinetic Studies

Kinetics and adsorption are the most important parameters in order to evaluate the adsorption dynamics of a process. For analyzing the controlling mechanism of the adsorption, pseudo-second-order kinetics model is used to test the experimental data of the adsorption of Cr (VI) ions by OBN and OKN.

Fig. 4.8 Langmuir isotherm for the reduction of chromium (VI) ion by organo-kaolin nanoparticle (OKN)

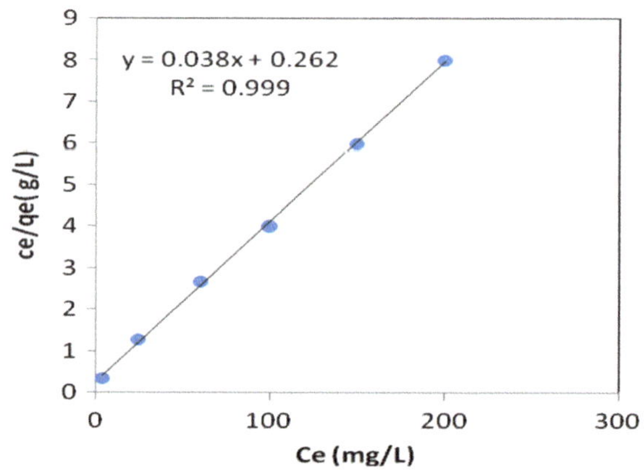

4.3.6.1 Pseudo-Second-Order Model

The adsorption kinetics can be described by pseudo-second-order equation. The linear form is the following:

$$\frac{t}{q_t} = \frac{1}{K_2 q_e^2} + \frac{1}{q_e} t \tag{4.4}$$

where t is the time (min), q_e and q_t are the amount of species adsorbed per unit mass of adsorbent (mg g $^{-1}$) at equilibrium and at any time t, respectively, and K_2 is the rate constant of pseudo-second-order adsorption (g mg^{-1} min^{-1}). The absorption of OKN and OBN nano-adsorbent is more appropriately described by pseudo-second-order kinetic model ($R^2 > 1$) was shown in Figs. 4.9 and 4.10, respectively.

Fig. 4.9 Pseudo-second-order kinetic reduction of Cr (VI) by OKN

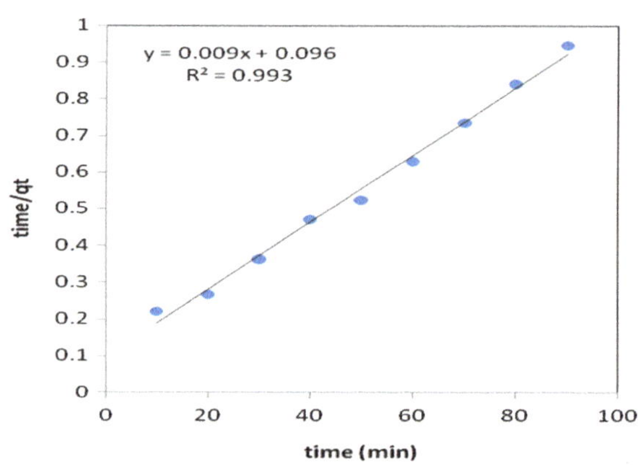

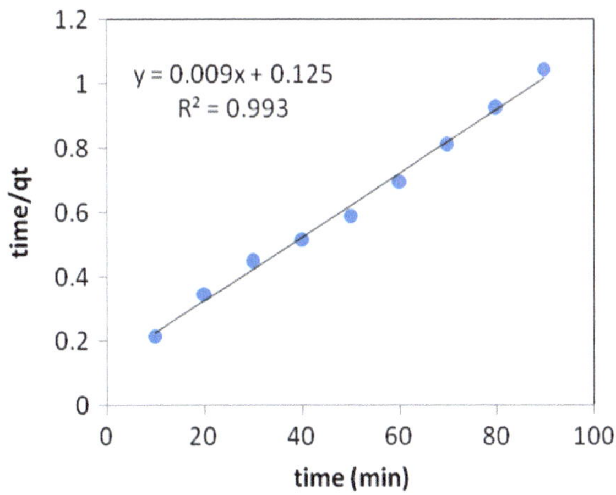

Fig. 4.10 Pseudo-second-order kinetic reduction of Cr (VI) by OBN

4.4 Conclusion

Adsorption studies were carried out using organo-kaolin nanoparticles and organo-bentonite nanoparticles. On comparing the results, it was concluded that organo-kaolin nanoparticles can be used as suitable and effective adsorbent for the reduction of Cr (VI) ions from wastewater. The optimum pH was found to be 2 for OKN and 3 for OBN. Similarly, the percentage reduction of metal ion was determined 95.3% for OKN and 85.2% for OBN. Equilibrium adsorption data for Cr (VI) were better fitted by Langmuir adsorption isotherm ($R^2 = 0.999$ for OKN and $R^2 = 0.994$ for OBN). The rate kinetics for was best fitted by pseudo-second-order model for both OKN and OBN. From the above results, it is concluded that organo-kaolin nanoparticles serve as a best and suitable adsorbent for the reducing the level of Cr (VI) from wastewater.

Acknowledgements The authors convey sincere thanks to the Management and Principal of Coimbatore Institute of Technology for support through the Technical Education Quality Improvement Programme (TEQIP) fund.

References

1. BKrishna, B.S., Mahadevaiah, N., Murty, D.S.R., Prakash, B.J.: Surfactant immobilized interlayer species bonded to montmorillonite as recyclable adsorbent for lead ions. J. Colloid Interfac. Sci. **271**(2), 270–276 (2004)
2. Wang, P., Du, M., Zhu, H., Bao, S., Yang, T., Zou, M.: Structure regulation of silica nanotubes and their adsorption behaviours for heavy metal ions: pH effect, kinetics, isotherms and mechanism. J. Hazardous Mater. **286**, 533–544 (2015)
3. Yagub, M.T., Sen, T.K., Afroze, S., Ang, H.M.: Dye and its removal from aqueous solution by adsorption: a review. Adv. Colloid Interfac. **209**, 172–184 (2014)

4. Crini, G.: Non-conventional low-cost adsorbent for dye removal: a review. Biores. Technol. **97**, 1061–1085 (2006)
5. Saleh, M.: On the removal of cationic surfactants from dilute streams by granular charcoal. Water Res. **40**, 1052–1060 (2006)
6. Abd El-Rahman, K.M., El-Sourougy, M.R.,Abdel-Monem, N.M.,Ismail, I.M.: Modelling the sorption kinetics of ceasium and strontiumions on zeolites. A. J. Nucl. Radiochem. Sci. **7**(2):21–27 (2006)
7. Krim, L.,Nacer, S.,Bilango, G.: Kinetics of chromium sorption on biomass fungi from aqueous solution. Amm. J. Environ. Sci. **2**(1), 31–36 (2006)
8. Zhao, F.,Repo, E., Sillanpaa, M., Meng, Y., Yin, D.,Tang, W.Z.: Green synthesis of magnetic EDTA—and/or DTPA—crossed—linked chitosin adsorbents for high effective removal of metals. Ind. Eng. Chem. Res. **54**(4) 1271–1281 (2015)
9. Atia, A.A.: Studies on the interaction of mercury (II) and uranyl(II) with modified chitosin resin. Hydrometallurgy **80**,13–22 (2005)
10. Zhou, Y., Zhou, W., Hou, D., Li, G., Wan, J., Feng, C., Tang, Z., Chen, S.: Metal-carbon hybrid electrocatalyst derived from ion-exchange resin containing heavy metals for efficient hydrogen evolution reaction. Small **12**(20), 2768–2774 (2016)
11. Nathaniel, E., Kurniawan, A., Soetaredjo, F.E., Ismadji, S.: Organo-bentonite for the adsorption of Pb(II) from aqueous solution: temperature dependent parameters of several adsorption equations. Desalination Water Treatment **36**, 1–9 (2006)
12. Qu, F., Zhu, L., Yang, K.: Adsorption behavoiur of volatile compound (VOCS) on porous clay heterogenous structures (PCH). J. Hazardous Mater. **170**, 7–12 (2009)
13. El, A.A.E.Z.A., Al-Hazmi, G.A.: Organically modified clay for adsorption of petroleum hydrocarbon. Eur. Chem. Bull. **4**(1–3), 87–91 (2015)
14. Maramis, V., Kurniawan, A., Ayucitra, A., Sunarso, J., Ismadji, S.: Removal of copper ions from aqueous solution by adsorption using LABORATORIES-modified bentonite (organo-bentonite). Front. Chem. Sci. Eng. **6**(1), 58–66 (2012)
15. Havva tutar Kahraman: Use of organo-montmorillonite nano clay as an environmentally friendly adsorbent for removal of hexavalent chromium. Proc. Int. Proc. Chem. Biol. Environ. Eng. **101**, 136–141 (2017)

Chapter 5
Impact of Sugar Industry Effluent on Quality of Groundwater at Rasulwadi—Sambarwadi, Maharashtra, India

P. V. Kumbhar and C. H. Wagh

Abstract Groundwater is the primary source of drinking and irrigation for both urban and rural areas. It is affected due to the discharge of improperly treated industrial effluents. Increasing the use of chemical fertilizers and pesticides in raising the sugarcane production has resulted in the degradation of groundwater quality and also affected the soil. The present study focuses on assessing the impact of sugar industry effluent on groundwater in and around Rasulwadi—Sambarwadi village where groundwater is used for domestic purpose. Groundwater samples were collected from ten different locations in the vicinity of Rasulwadi—Sambarwadi village and analyzed for physico-chemical parameters such as pH, electrical conductivity, total dissolved solids, dissolved oxygen, total alkalinity, total hardness, chloride, nitrate, calcium, magnesium, sodium and potassium. Also, irrigation suitability was checked using SAR, RSC, Na% and PI. Water quality index was calculated to assess the level of pollution and deterioration of groundwater quality in both villages. Chlorides and hardness seem to be exceeding permissible limits as per the BIS. SAR, RSC and Na% were observed within the limits. It is found that all water samples are unfit for drinking purpose but suitable for irrigation purposes. The investigation suggests that water quality management is an important issue for the sustenance of human civilization and must become a major priority.

Keywords Groundwater quality · Physico-chemical characteristics · Sugar industry effluent · Water quality index

5.1 Introduction

Water is one of the most important resources found on earth. It is an essential requirement not only for human life but also for industrial development. Water is available in the two forms as surface water and groundwater. Groundwater is the part of the hydrological cycle. Groundwater is the major primary source of drinking water and

P. V. Kumbhar (✉) · C. H. Wagh
Department of Civil Engineering, Walchand College of Engineering, Sangli, India
e-mail: prachikhumbhar12@gmail.com

© Springer Nature Singapore Pte Ltd. 2020

V. Sivasubramanian and S. Subramanian (eds.), *Global Challenges in Energy and Environment*, Lecture Notes on Multidisciplinary Industrial Engineering, https://doi.org/10.1007/978-981-13-9213-9_5

also used for irrigation purpose in India. Rapidly, industrialization, growing population in developing countries like India, has affected the availability and quality of groundwater due to the improper waste disposal. According to the WHO organization, about 80% of all the diseases in human beings are caused by water [1].

The municipal solid waste and industrial wastewater are the most important causes of groundwater pollution. Municipal solid waste and industrial wastewater come in contact with natural water bodies. The wastewater contains a large amount of microorganisms and chemical constituents. In small-scale industries and large-scale industries, they release wastewater with or partial treatment on the land surface or into natural water bodies. This discharge of industrial effluent percolates through the soil which causes the contamination of groundwater, surface water and soil.

Sugar industry is one of the most important agro-based industries in India. Sugar industry consumes a large volume of water and chemicals in all different processing units. Sugar industry effluent consists of all types of toxic chemicals, heavy metals, suspended solids, sludge, press mud and bagasse. The effluent comings from industries sometimes percolate through the subsoil and reach the groundwater table forming contaminated pool, which disturb the natural groundwater quality by changing its chemical composition [2].

The present study has been carried out to assess the water quality parameters of groundwater and soil in the field of agricultural areas.

5.2 Materials and Methods

5.2.1 Study Area

The place of study is Shree Vasantdada Patil Shetkari Sahakari Sakhar Karkhana which is located at Sangli district of Maharashtra. From the boiling house, mill house and manufacturing process, the total effluent 370 m^3/day transfers to the effluent treatment plant, and then this treated effluent is discharged into the agricultural area in Rasulwadi—Sambarwadi village which is located 20 km away from the sugar industry. It geographically lies between 16° 55′12.13″ N latitude and 74° 39′32.65″ E longitude for Rasulwadi and 16° 56′5.42″ N latitude and 74° 39′53.74″ E for Sambarwadi.

5.2.2 Groundwater Sampling and Analysis

The present study was conducted to analyze the effect of sugar industry effluent on the quality of groundwater at Rasulwadi—Sambarwadi village. Groundwater samples were collected from ten different locations in the vicinity of Rasulwadi—Sambarwadi villages. One liter capacity of clean polythene plastic bottle has been used to collect the bore well and open well samples. The bottles were tightly sealed after collection and labeled in the field. Then, various physico-chemical parameters such as pH, EC, TDS, hardness, chloride, alkalinity, carbonate, bicarbonate, calcium, magnesium, sodium and potassium were analyzed by using a standard method. Also, agricultural soil was collected from Rasulwadi—Sambarwadi villages and the parameter such as SAR, RSC, Na% and PI were analyzed.

Water quality index (WQI) is the index number which represents the overall water quality for any future application at a certain location and time. The objective of water quality index is to turn complex water quality data into detailed information useful for public [3]. Water quality index (WQI) is valuable and unique rating to depict the overall water quality status in a single term that is helpful for the selection of appropriate treatment technique to meet the concerned issues [4]. It is one of the most effective and simple tool to assess the quality of water for suitability of various purposes.

For the calculation of water quality index, important water quality parameters were chosen. This water quality index (WQI) is calculated by using the drinking water quality standards recommended by Bureau of Indian Standards (BIS), Indian Council of Medical Research (ICMR) and World Health Organization (WHO).

The weighted arithmetic index method has been used for the calculation of water quality index (WQI). Then, calculated subindex and quality rating (qn) by using the following expression are as follows:

$$qn = (Cn/Sn) * 100 \tag{5.1}$$

where

qn Quality rating or subindex for the *n*th water quality parameter.
Cn Estimated concentration of the *n*th parameter at a sampling station
Sn Standard permissible value of *n*th parameter

The unit weight (Wn) for each water quality parameter is calculated by using the following formula:

The overall water quality index is calculated by aggregating the quality rating with the unit weight linearly (Table 5.1).

Table 5.1 Rating of water quality

WQI	Rating of water quality	Grade
0–25	Excellent water quality	A
26–50	Good water quality	B
50–75	Poor water quality	C
76–100	Very poor water quality	D
>100	Unsuitable for drinking purpose	E

Source Jena et al. [6]

$$WQI = \sum qn * Wn / \sum Wn \quad (5.2)$$

5.3 Results and Discussion

Characteristics of Groundwater.

5.3.1 Physico-Chemical Characteristics of Groundwater Samples at Sambarwadi

In this study, pH of groundwater samples ranges from 6.95 to 7.57. The value of all site indicates that pH value within the desirable limit and which is slightly alkaline in nature may be due to discharge of industrial effluent. Figure 5.1 shows the EC of all location ranges from 1.85 to 3.57 mS/cm. Electrical conductivity is a measure of the concentration of ionized substances in water [5]. Electrical conductivity values were found in higher than desirable limit. So, impact of industry effluent on groundwater quality by increasing EC.

Figure 5.2 shows the concentration of TDS and alkalinity of groundwater samples. TDS of groundwater samples ranges from 904 to 1815 mg/l. The value of TDS of sample exceeded the prescribed limit given by WHO except SB4 groundwater sample. The TDS values indicate not only the impact of sugar industry on the quality of groundwater but also on other factors involve for causes concern on the surrounding environment.

Alkalinity is measuring the acid neutralizing capacity of water. Desired limit of alkalinity is 200 mg/l and permissible limit of alkalinity is 600 mg/l in the absence of alternate water source according to BIS 2012.

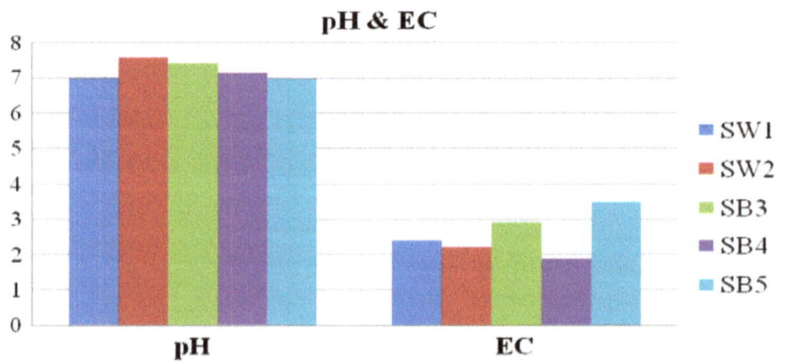

Fig. 5.1 Variation of pH and EC

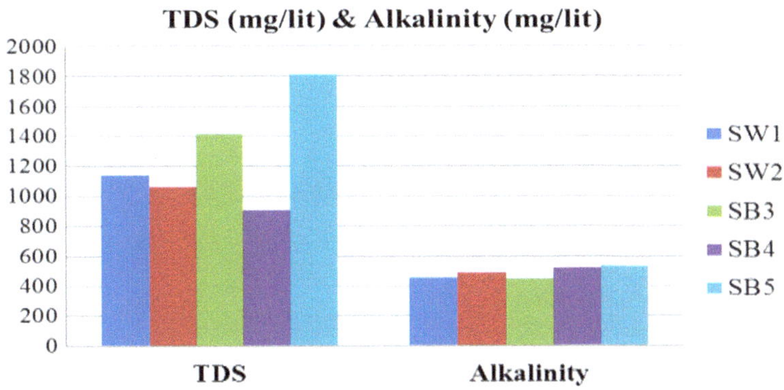

Fig. 5.2 Variation of TDS and alkalinity

Alkalinity of groundwater samples ranges from 450.4 to 534.4 mg/l. This value indicates the alkalinity was higher than desired limit.

Figure 5.3 shows the data obtained of hardness and chloride in mg/l. Hardness contains bicarbonate, chloride and sulfates of calcium and magnesium. The total hardness is due to excessive use of lime in sugar processing. Hardness of groundwater ranges from 400 to 520.4 mg/l. The total hardness values higher than desired limit given by BIS and within the permissible limit are given by WHO.

Chloride is one common anionic constituent of water whose high concentration imparts a salty taste to water making it unacceptable for drinking [5]. Desired limit of chloride is 250 mg/l, and permissible limit in the absence of alternate source is 1000 mg/l according to BIS 2012. In this study, value ranges from 399.6 to 495.2 mg/l. This value indicates exceeding the desired limit according to BIS.

The calcium data obtained is presented in Fig. 5.4. The permissible limit of calcium in drinking water is 75 mg/l. Calcium concentration of groundwater samples is 124 to 151.2 mg/l. Calcium in the groundwater sample exceeded the permissible limit of drinking standards indicating the impact of effluent on quality of groundwater.

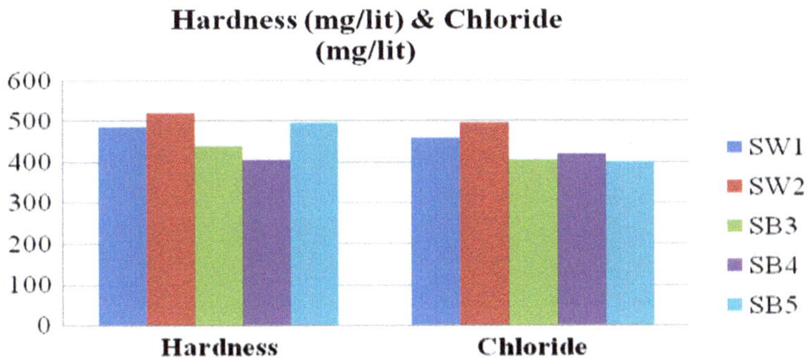

Fig. 5.3 Variation of hardness and chloride

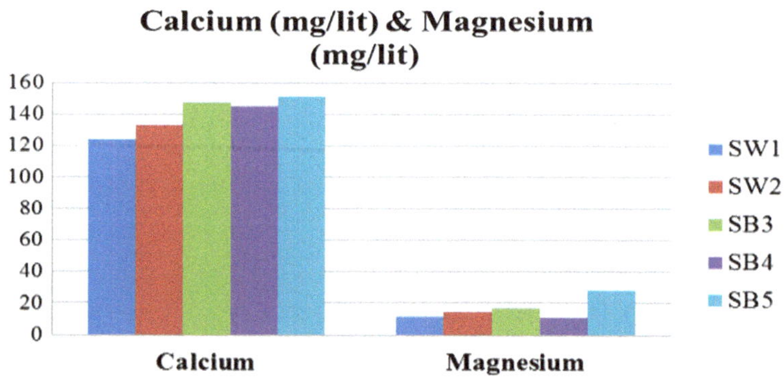

Fig. 5.4 Variation of calcium and magnesium

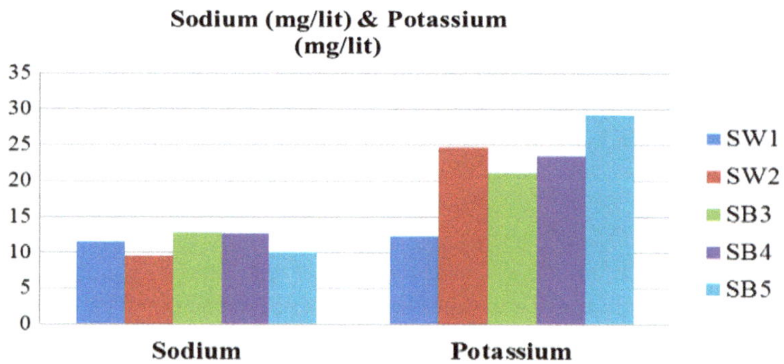

Fig. 5.5 Variation of sodium and potassium

The magnesium data obtained is presented in Fig. 5.4. The permissible limit of magnesium in drinking water is 30 mg/l. Magnesium concentration in groundwater ranges from 10.64 to 28.6 mg/l is within the permissible limit.

Figure 5.5 shows the variation of sodium and potassium in mg/l. According to WHO, the permissible limit of sodium is 200 mg/l. The concentration of sodium ranges from 9.5 to 12.82 mg/l, and this is below the permissible limit. The potassium ranges from 12.25 to 29.06 mg/l.

Figure 5.6 shows the variation of DO and nitrate at different locations. DO is a very important parameter for aquatic life. In this study, DO values of all samples were found above the permissible limit 5 mg/l.

According to the drinking water quality standards, the permissible limit of nitrate is 45 mg/l. The nitrate ranges from 15 to 30 mg/l. This value indicates that nitrate value is below the permissible limit.

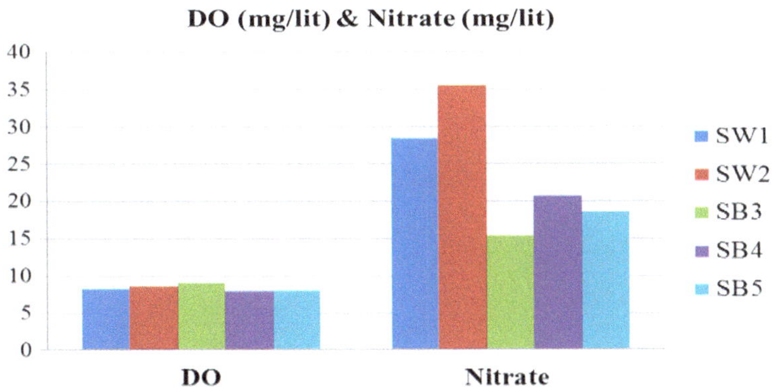

Fig. 5.6 Variation of DO and nitrate

5.3.2 Physico-Chemical Characteristics of Groundwater Samples at Rasulwadi

In this study, pH of groundwater samples ranges from 6.63 to 7.2. The value of all site indicates that pH value within the desirable limit and which is slightly alkaline in nature may be due to discharge of industrial effluent. Figure 5.7 shows the EC of all location ranges from 2.62 to 7.26 mS/cm. Electrical conductivity is a measure of the concentration of ionized substances in water [5]. Electrical conductivity values were found in higher than desirable limit. So, impact of industry effluent on groundwater quality by enhancing EC.

Figure 5.8 shows the concentration of TDS and alkalinity of groundwater samples. TDS of groundwater samples ranges from 1359 to 3950 mg/l. The value of TDS of sample exceeded the prescribed limit given by WHO except SB4 groundwater sample. The TDS values indicate not only the impact of sugar industry on the quality

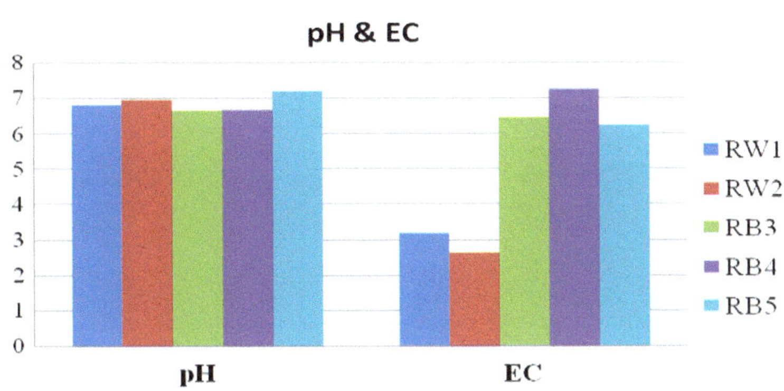

Fig. 5.7 Variation of pH and EC

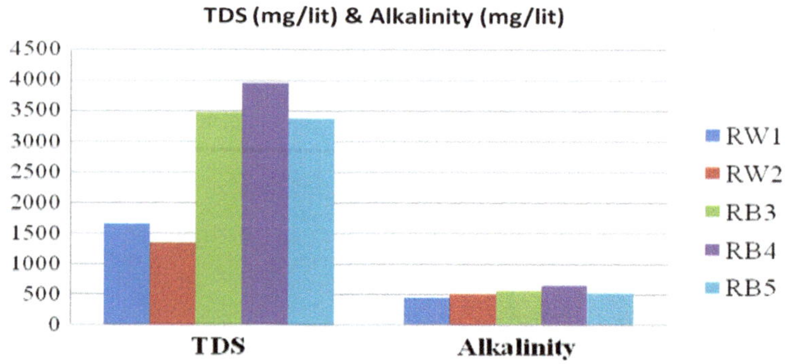

Fig. 5.8 Variation of TDS and alkalinity

of groundwater but also on other factors involve for causes concern on the surrounding environment.

Alkalinity is measuring the acid neutralizing capacity of water. Desired limit of alkalinity is 200 mg/l and permissible limit of alkalinity is 600 mg/l in the absence of alternate water source according to BIS 2012. Alkalinity of groundwater samples ranges from 448 to 648 mg/l. This value indicates the alkalinity was higher than desired limit.

Figure 5.9 shows the data obtained of hardness and chloride in mg/l. Hardness contains bicarbonate, chloride and sulfates of calcium and magnesium. The total hardness is due to excessive use of lime in sugar processing. Hardness of groundwater ranges from 378 to 476 mg/l. The total hardness values higher than desired limit given by BIS and within the permissible limit are given by WHO.

Chloride is one common anionic constituent of water whose high concentration imparts a salty taste to water making it unacceptable for drinking [5]. Desired limit of chloride is 250 mg/l, and permissible limit in the absence of alternate source is 1000 mg/l according to BIS 2012. In this study, value ranges from 379.88 to 491.85 mg/l. This value indicates exceeding the desired limit according to BIS.

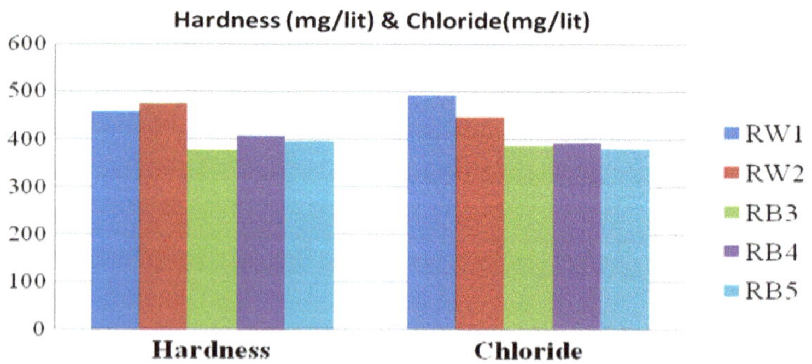

Fig. 5.9 Variation of hardness and chloride

The calcium data obtained is presented in Fig. 5.4. The permissible limit of calcium in drinking water is 75 mg/l. Calcium concentration of groundwater samples is 136.14 to 153.76 mg/l. Calcium in the groundwater sample exceeded the permissible limit of drinking standards indicating the impact of effluent on quality of groundwater.

The magnesium data obtained is presented in Fig. 5.10. The permissible limit of magnesium in drinking water is 30 mg/l. Magnesium concentration in groundwater ranges from 5.82 to 26.315 mg/l is within permissible limit.

Figure 5.11 shows the variation of sodium and potassium in mg/l. According to WHO, the permissible limit of sodium is 200 mg/l. The concentration of sodium ranges from 12.7 to 52.5 mg/l, and this is below the permissible limit. The potassium ranges from 11.9 to 56.3 mg/l.

Figure 5.12 shows the variation of DO and nitrate at different locations. DO is a very important parameter for aquatic life. In this study, the DO values of all samples were found above the permissible limit 5 mg/l.

According to the drinking water quality standards, the permissible limit of nitrate is 45 mg/l. The nitrate ranges from 19.6 to 36.2 mg/l. This value indicates that nitrate value is below the permissible limit.

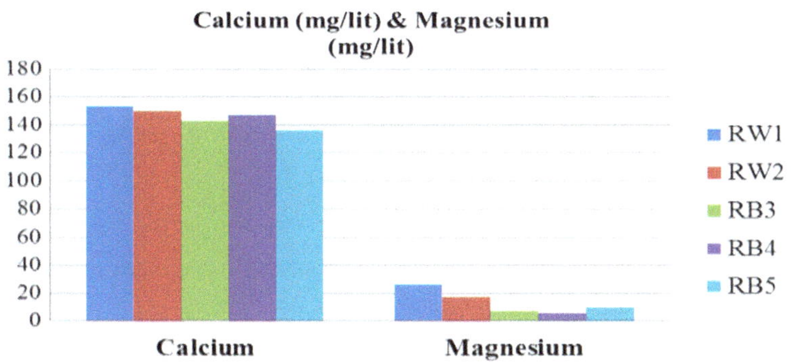

Fig. 5.10 Variation of calcium and magnesium

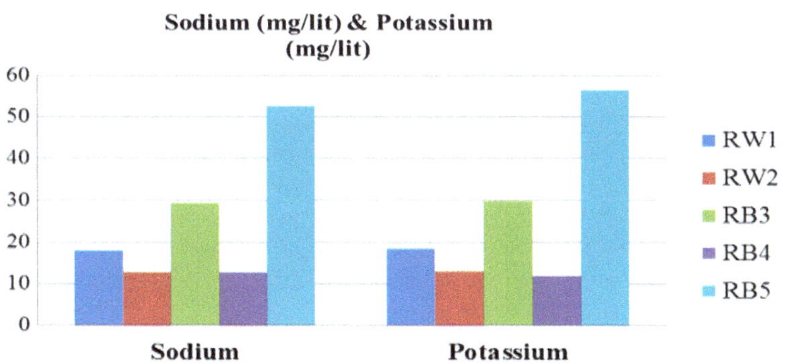

Fig. 5.11 Variation of sodium and potassium

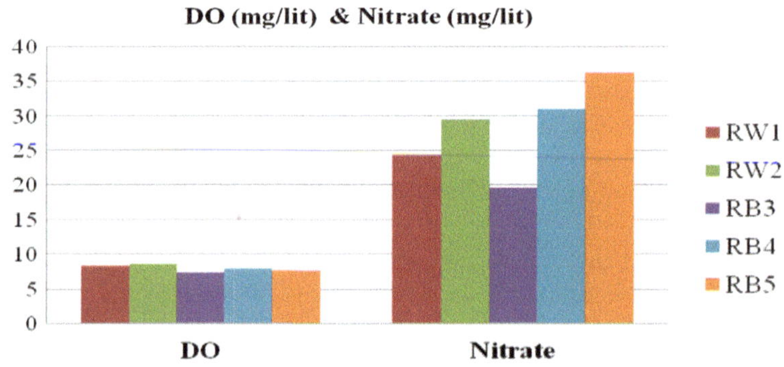

Fig. 5.12 Variation of DO and nitrate

Table 5.2 Characteristics of soil

Parameter	Sambarwadi	Rasulwadi
pH	7.85	7.96
EC (ms/cm)	0.433	0.478
SAR	2.526	1.809
RSC	−0.2	−0.18
Na%	57.70	48.23
PI	78.27	70.93

5.3.2.1 Soil Characteristics

Table 5.2 shows the characteristics of soil at Rasulwadi—Sambarwadi agricultural area. In the present study, all soil characteristics parameters were within standard limit.

5.3.2.2 Water Quality Index (WQI)

See (Table 5.3).

5.4 Conclusion

1. On the basis of above discussion, it is found that the groundwater in the Rasulwadi—Sambarwadi village was contaminated due to the higher concentration of chlorides, hardness, electrical conductivity and total dissolved solids which was higher than drinking water quality standards BIS and WHO.
2. TDS of groundwater sample was higher for majority of sample which indicates that quality of groundwater is not suitable for drinking purpose.

3. SAR, RSC, Na% and PI of soil characteristics were within the standard limit. Hence, this groundwater is suitable for irrigation purpose.
4. WQI of all different locations ranges from 140.538 to 236.993. All WQI of the sample was exceeded 100, standard limit for drinking water quality. It is concluded that groundwater sample quality is a poor water quality.
5. The analysis reveals that the groundwater of the area needs some water treatment before consumption and needs periodic testing of water quality.

References

1. Shivaprasad, H., Nagarajappa, D.P., Sham Sundar, K.M.: A study on physico-chemical characteristics of borewell water in sugar town, Mandya City, Karnataka State, India. Int. J. Eng. Res. Appl. **4**(7)(Version 1), 112–123 (2014)
2. Agale, M.C., Patel, N.G., Patil, A.G.: Impact of sugar industry effluents on the quality of ground water from Dahiwad Village, Dist-Dhule(M.S.). Arch. Appl. Sci. Res. **5**(2), 58–60 (2013)
3. Qureshimatva, U.M., Gamit, S.B., Patel, R.D., Solanki, H.A.: Determination of physico-chemical parameters and water quality index (Wqi) of Chandlodia Lake, Ahmedabad, Gujarat, India. J. Environ. Anal. Toxicol. **5**(4) (2015)
4. Tyagi, S., Sharma, B., Singh, P., Dobhal, R.: Water quality assessment in terms of water quality index. Am. J. Water Res. **1**(3), 34–38 (2013)
5. Agarwal, A., Deswal, S.: Assessment of ground water quality in vicinity of industries in Bijnor, U.P, India. Int. J. Emerg. Technol. **8**(1):745–750 (2017)
6. Jena, V., Dixit, S., Gupta, S.: Assessment of water quality index of industrial area surface water samples. Int. J. ChemTech Res. **5**(1), 278–283 (2013)
7. APHA.: Standard methods for the examination of water and wastewater, 21th edn. American Public Health Association, Washington DC (2005)
8. WHO.: Guidelines for drinking-water quality, health criteria and other supporting information, V. 2, 2nd edn, pp. 940–949. World Health Organisation, Geneva (1996)
9. BIS.: Indian standards specifications for drinking water. Bureau of Indian Standards, ISO:10500 (1994)

Table 5.3 Water quality index

Site	WQI	Water quality status
SW1	148.279	Poor water quality
SW2	155.619	Poor water quality
SB3	154.345	Poor water quality
SB4	140.538	Poor water quality
SB5	173.714	Poor water quality
RW1	169.803	Poor water quality
RW2	162.359	Poor water quality
RB3	212.645	Poor water quality
RB4	236.993	Poor water quality
RB5	214.809	Poor water quality

10. Srinivas, J., Purushotham, A.V., Murali Krishna, K.V.S.G.: Determination of water quality index in industrial areas of Kakinada, Andhra Pradesh, India. Int. Res. J. Environ. Sci. **2**(5), 37–45 (2013)
11. Yadav, E., Wahane, J.K., Choubey, O.N.: Impact of sugar industry effluents on the quality of groundwater Near Bankhedi sugar industry Dist. Narsinghpur (Mp) India. J. Ind. Pollut. Control (2015)

Chapter 6
Influence of Chloride Content and Exposure Time on Corrosion Behavior of AZ80 Wrought Mg Alloy

Gopal D. Gote, Gajanan M. Naik and S. Narendranath

Abstract This study aims to investigate the corrosion behavior of wrought AZ80 magnesium alloys in different chloride ion concentrations and exposure time. During the study, the effect of exposure time and chloride content on the corrosion resistance of AZ80 wrought Mg alloy has been studied with 2, 3.5, and 5 wt% NaCl aqueous solution for 12 h and 24 h exposure time. Charge transfer resistance for each sample was established using the Nyquist plot and corrosion rate obtained from polarization curve by adopting Tafel extrapolation method. Corrosion morphology was examined using scanning electron microscopy and XRD. The study revealed that corrosion product layer formed at initial stage was observed unstable after short exposure time which results in decrease in corrosion resistance at initial stage. An increase of chloride content in aqueous environment reduces the corrosion resistance of AZ80 wrought Mg alloy.

Keywords AZ80 · Corrosion · Nyquist plot · Polarization curve

6.1 Introduction

Applications of magnesium alloys are increasing day by day because of its low density, high specific strength, and good machinability [1]. Mg alloys have great potential to replace aluminum and steel due to its lightweight. Also, the use of magnesium in automobile and aerospace can reduce the fuel consumption and harmful gases

G. D. Gote
Department of Mechanical Engineering, IITB, Mumbai 400076, Maharashtra, India
e-mail: gopalgote99@gmail.com

G. M. Naik (✉) · S. Narendranath
Department of Mechanical Engineering, National Institute of Technology Karnataka, Surathkal, Mangaluru 575025, Karnataka, India
e-mail: gajamnaik@gmail.com

G. M. Naik
Department of Mechanical Engineering, Mangalore Institute of Technology and Engineering, Moodbidri, Mangaluru 574225, Karnataka, India

© Springer Nature Singapore Pte Ltd. 2020

V. Sivasubramanian and S. Subramanian (eds.), *Global Challenges in Energy and Environment*, Lecture Notes on Multidisciplinary Industrial Engineering, https://doi.org/10.1007/978-981-13-9213-9_6

emission [1–3]; along with this, Mg alloys were used in biomedical field because of its biodegradable properties. But magnesium has inadequate applications in marine environment because of its poor corrosion resistance. Currently, Mg alloys used as sacrificial anode to protect ship hull and offshore structures in marine field. Corrosion of magnesium alloys is now emerging topic [4]. Especially in case of AZ magnesium alloy, galvanic corrosion is more detrimental due to the presence of secondary particles ($Mg_{17}Al_{12}$), and it is serious issue from last few years. Corrosion on magnesium will be localized in nature and it is distributed over the surface, further shallow pits are observed; instead of deeper pits, it became wider and cover more surface; most favourable reason for that is the production of OH^- ions which is the by-product of cathodic reaction and because of these ions, localized pH get increased, also magnesium hydroxide film gets stabilized; therefore it acts as barrier for further corrosion [1]. A Pardo et al. suggested that AZ80 Mg alloys have greater corrosion resistance than pure Mg and AZ90. Also, from study, it was shown that AZ80 alloy has lower corrosion resistance; when it is immersed for 7 days further, AZ80 alloys showed better corrosion resistance for 28 days of immersion [5]. Zhao et al. observed there is a significant impact of chlorine ion concentration on ZE41 magnesium alloy, and it was revealed that corrosion rate increased with increase of chlorine ion concentration [6].

Immersion method, electrochemical corrosion test, and hydrogen evaluation test are known for corrosion behavior study. Among all, electrochemical corrosion test is given the maximum priority because it provides provision of continuous corrosion monitoring. Along with this, corrosion current density can be obtained ($i_{\mathbf{corr}}$) which can be used for calculating corrosion rate. Corrosion current density is directly proportional to the corrosion rate and has relation according to ASTM G59 [7] as follows:

$$\mathrm{CR(mm/yr)} = 3.27 \times 10^{-3} \frac{i_{Corr} \times A}{\rho}$$

where CR is the corrosion rate in mm/yr, $i_{\mathbf{Corr}}$ is the corrosion current density in $\mu A/cm^2$, A is the equivalent weight 24.3 gm/mol, and ρ is the density of the Mg alloy 1.74 g/cc.

AZ80 magnesium alloy is the promising material used in the aerospace alloys; it has better mechanical properties than other AZ alloys. Corrosion of AZ Mg alloys is dependent on the alloying elements, distribution of secondary phases, and environment. Among these variables, corrosion media or environment play a significant change in the corrosion behavior of magnesium and its alloys.

Following reaction represents the corrosion process in the aqueous chlorinated media and formation of corrosion product, which consists of $Mg(OH)_2$ and $Mg(Cl)_2$. $Mg(OH)_2$ produce a stable corrosion passive layer, but $Mg(Cl)_2$ dissolve in the environment [4, 6].

$$\mathrm{Mg} \rightarrow \mathrm{Mg}^{2+} + 2\mathrm{e}^{-} \tag{6.1}$$

$$2H_2O + 2e^- \rightarrow H2 + 2(OH)^- \tag{6.2}$$

$$Mg^{2+} + 2(OH)^- \rightarrow Mg(OH)_2 \tag{6.3}$$

$$Mg^{2+} + 2Cl^- \rightarrow Mg(Cl)_2 \tag{6.4}$$

In case of AZ80 Mg alloy, lack of literature available on formation of stable $Mg(OH)_2$ passive layer in short time exposure during corrosion study. Therefore, in this study, corrosion behavior of AZ80 for the short time exposure has been studied to understand the corrosion behavior at the initial stage of the formation of passive layer in different concentrations of NaCl solution.

6.2 Experiment

AZ80 wrought magnesium alloy having diameter 16 and 4 mm thick is prepared for the electrochemical corrosion study. Samples are grounded to cloth polishing followed by polishing with emery paper up to 2000 grade. Further, polished surface has been cleaned using ethanol before immersing into chloride media. The different chlorine ion solutions are prepared with 2, 3.5, and 5 wt% NaCl. Samples are kept for 12 and 24 h duration; then, electrochemical test has performed. Electrochemical test was carried using electrochemical cell supplied by ACM three electrode-type electrochemical setups. AZ80 Mg alloys act as working electrode. Calomel electrode and graphite are used as reference and auxiliary electrode, respectively. Polarization study has been done using 60 mV/min sweep rate in the potential range of −250 to 250 mV; along with this, electrochemical impedance spectroscopy test has been carried out in the frequency range of 10,000–0.001 Hz. Corrosion rate has been calculated using corrosion current density which is obtained from polarization curve by adopting the Tafel extrapolation method.

After electrochemical study, corrosion morphology has been observed by scanning electron microscopy. Further, XRD analysis was carried out to study the chemical composition of corrosion product.

6.3 Results and Discussion

6.3.1 Electrochemical Impedance Spectroscopy

Figure 6.1b, d, and f represents the Nyquist plot for AZ80 Mg alloy after 0, 12, and 24 h immersion in 2, 3.5, and 5 wt% NaCl solution. The Impedance spectra plot revealed the same trend at different concentrations of NaCl solution. Study found

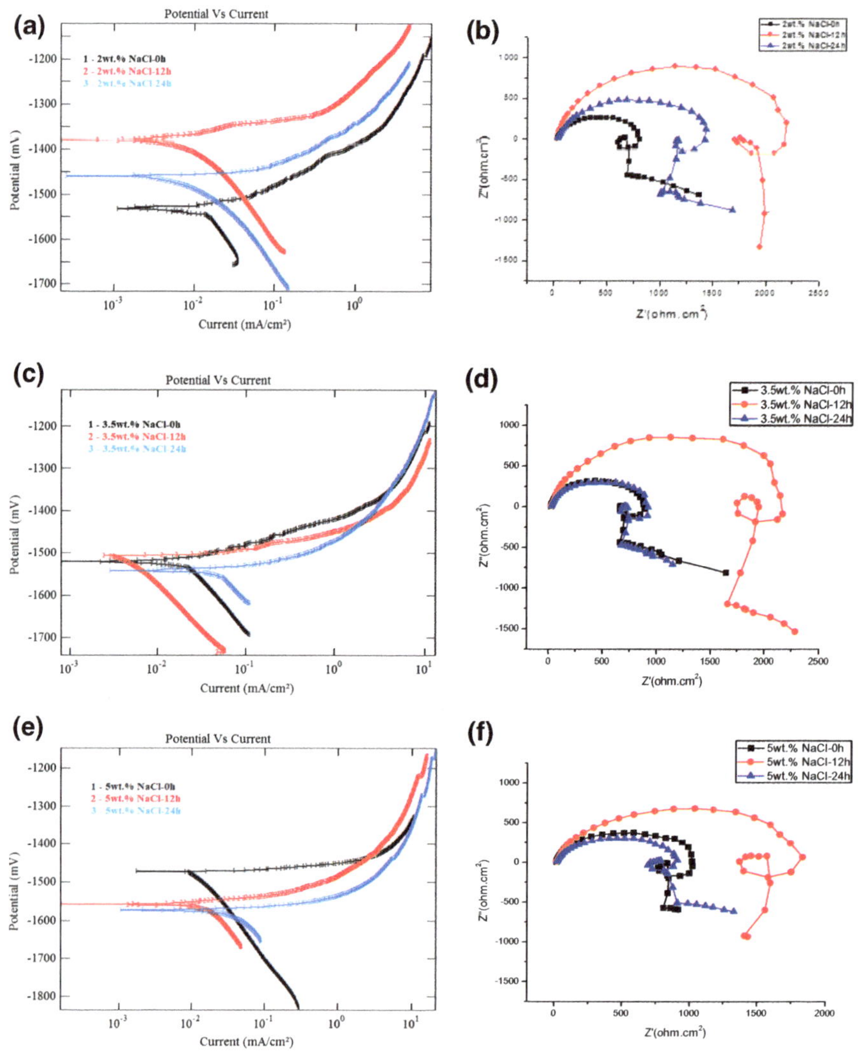

Fig. 6.1 **a**, **c**, and **e** represents polarization curves for AZ80 Mg alloy in 2, 3.5, and 5% wt NaCl, respectively. **b**, **d,** and **f** represents Nyquist plot for AZ80 Mg alloy in 2, 3.5, and 5% wt NaCl, respectively

that irrespective of NaCl concentration, AZ80 alloy shows increased charge transfer resistance for 12 h immersion, but corrosion resistance decreased for 24 h immersion. This clearly shown that corrosion product layer forms in 12 h immersion, but further immersion leads to dissolve protective layer. This clearly indicates that unstable corrosion layer formed during short time exposure because of low protective oxide layer which does not have potential to impede the corrosion attack and also possibility of broken and uneven hydroxide layer which provides the path for chloride media to reach base metal [5, 8]. Figure 6.2 shown the SEM morphologies of AZ80 Mg alloy, from Fig. 6.2 it was observed that severe damage on hydroxide layer after 24 h immersion. Similar observation was made by Pardo [5]. Table 6.1 indicates the charge transfer resistance values which were calculated from the capacitive arc; fresh sample in 2 wt% NaCl shows larger charge transfers resistance (2315 Ω cm^2) than 3.5 and 5 wt% NaCl solution. This is due to corrosion activity in the 2 wt% NaCl which is lower compared to the 3 and 5 wt% NaCl solution. This study reveals that AZ80 wrought Mg alloy shows the capacitive and inductive behavior; similar observation has been done for AZ91 and cast Mg alloy for 24 h immersion by Cao

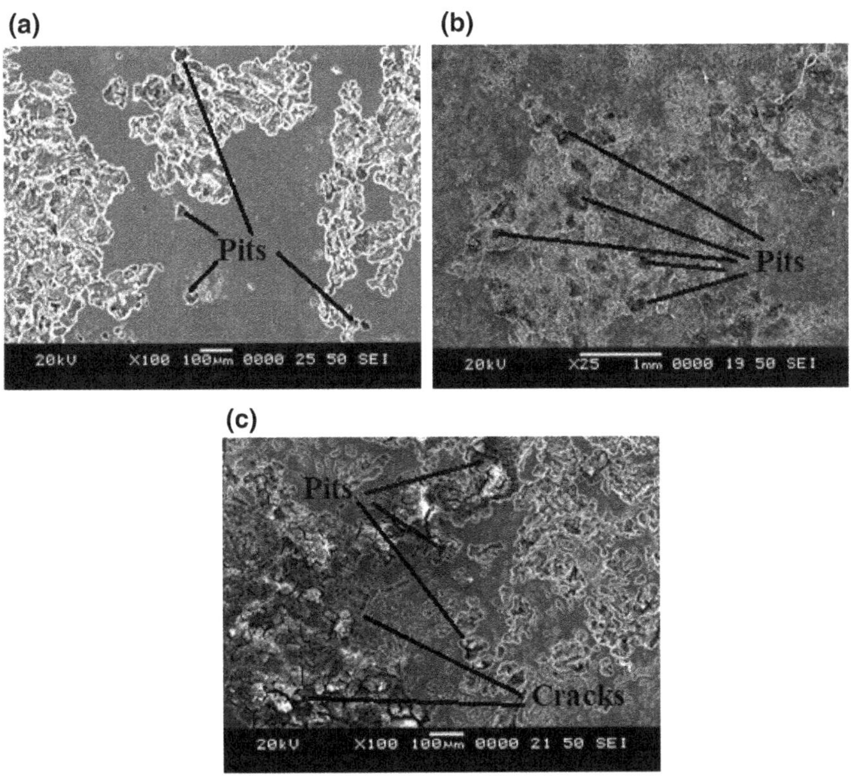

Fig. 6.2 Corrosion morphology of wrought AZ80 Mg alloy in 3.5% wt NaCl after electrochemical corrosion test for **a** 0 h immersion, **b** 12 h immersion, and **c** 24 h immersion

Table 6.1 Values of corrosion current density, corrosion potential, charge transfer resistance, and estimated corrosion rate

Sample type	R_{ct} (Ω cm^2)	E_{Corr} (mV)	I_{corr} (mA/cm^2)	Corrosion rate (mm/yr)
0 h exposed in 2% NaCl	2315	−1531.2	0.0051276	0.234
12 h exposed in 2% NaCl	816.9	−1380.3	0.0035036	0.1599
24 h exposed in 2% NaCl	1531	−1461.5	0.005012	0.2288
0 h exposed in 3.5% NaCl	1059	−1521.3	0.028436	1.2986
12 h exposed in 3.5% NaCl	2485	−1508.3	0.00986	0.4503
24 h exposed in 3.5% NaCl	982.3	−1543.5	0.053026	2.4215
0 h exposed in 5% NaCl	1034	−1471.8	0.03168	1.4467
12 h exposed in 5% NaCl	1956	−1559.3	0.017441	0.7965
24 h exposed in 5% NaCl	931.3	−1573.6	0.08862	4.047

et al. [9]. But Pardo et al. [10] did not observe inductive arc for AZ80 Mg alloy for long time exposure; this might because of exposure conditions. Occurrence of inductive loop signifies the pitting corrosion discussed by Jiang et al. [12].

6.3.2 Potentiodynamic Polarization Study

Figure 6.1a, c, and e indicates the potentiodynamic polarization curves for AZ80 wrought Mg alloys after short time immersion in 2, 3.5, and 5 wt% NaCl solution. From polarization curve, corrosion current densities for each sample have been measured by adopting the Tafel extrapolation method. Table 6.1 presents the current density and corrosion rate (mm/yr) for AZ 80 alloys. The corrosion rate was estimated using Eq. (6.1) [7]. Polarization curves for sample immersed in 2, 3.5, and 5 wt% show similar trend. Here, corrosion current density of AZ80 Mg alloy slightly shifted toward left for 12 h immersed sample irrespective of chloride concentration. This is due to the formation of passive layer, which arrests the ion movement during corrosion study. Therefore, corrosion rate for 12 h immersed sample is lower than the fresh sample. Further, corrosion current density increases for 24 h immersed sample than 12 h immersed sample irrespective of chloride content; the same finding has been observed in Pardo et al. [11]; it clearly shows that passive layer of oxide is unstable due to discontinuous layer as mentioned by Pardo [5]. Therefore, corrosion attack is severe in 24 h immersed sample. The corrosion potential (E_{corr}) during immersion of 2 wt% shows significant change with immersion up to 12 h shifted to noble side but again shifted down for 24 h immersion; this also gives the reason for unstable layer of corrosion product. The detailed explanation about unstable layer is given in the corrosion morphology.

6.3.3 Corrosion Morphology

Corrosion morphology was studied for sample immersed in 3.5 wt% NaCl solution for 0, 12, and 24 h immersion as shown in Fig. 6.3. From Fig. 6.3 observed pits over the surface of AZ80 Mg alloys, further these pits became wider it covers entire area of the specimen. Fresh sample surface observed more corrosion attack when compared to 12 h and 24 h immersed sample because there is no adherence for protection. It shows deeper pits, and it propagating severely. Fig. 6.2b shows that after 12 h immersion, corrosion layer is formed throughout the surface, and in some extent, it slows down the corrosion attack; this shows good agreement with corrosion current density which is mentioned in Table 6.1. Form Fig. 6.2c, it is observed that corrosion product layer is severely damaged because of localized corrosion. It is observed that for 24 h immersion sample corrosion layer is not protective and obtained higher corrosion current density than 12 h immersed sample. It shows that corrosion product layer for AZ80 Mg alloy is not stable for short time exposure.

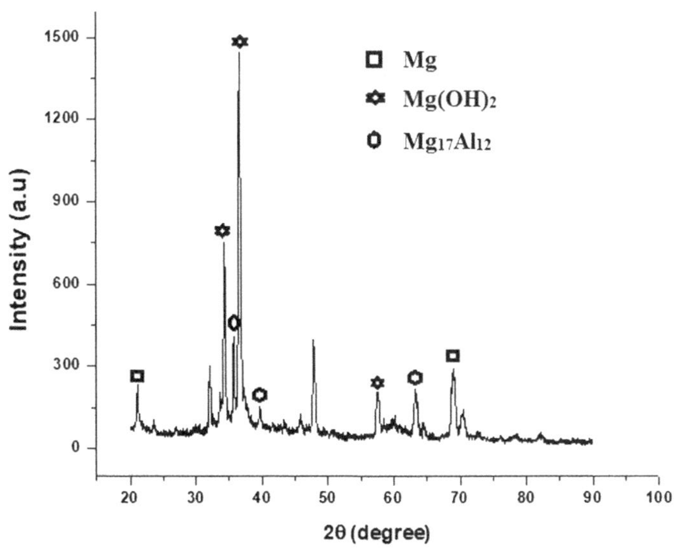

Fig. 6.3 XRD Analysis of AZ80 wrought Mg alloy after 24 h immersion in 5 wt% NaCl

6.3.4 XRD Result

Figure 6.3 presents the XRD pattern for corroded AZ80 Mg alloy which was immersed for 24 h in 5 wt% NaCl solution. The high-intensity peaks of $Mg(OH)_2$ were observed in the corroded sample. This signifies the $Mg(OH)_2$ is the main constituent in the corrosion product. Along with this, Mg and $Mg_{17}Al_{12}$ (β particle) secondary phase has been observed during the study. Here, α Mg (anodic phase) dissolved due to inter galvanic corrosion between α Mg and β particles, and β particles are fallen down or separated from the matrix [1, 10].

6.4 Conclusion

1. The short time immersion of AZ80 alloys yields unstable passive layer which shows decreased corrosion resistance. This means 24 h immersion is not sufficient time to get stable passive layer irrespective of the concentration of NaCl solution.
2. Corrosion rate of AZ80 wrought Mg alloy significantly increases with increase of chloride content.
3. AZ80 wrought Mg alloy shows capacitive and inductive behavior in the chloride media for short time exposure which conforms the pitting corrosion.
4. XRD study reveals the presence of $Mg_{17}Al_{12}$ secondary particle which acts as cathodic sites in Mg matrix which indicates severe galvanic corrosion.

Acknowledgements This work was supported by DRDO-NRB, Government of India, under grant number NRB/4003/PG/366.

References

1. Cao, F., Song, G., Atrens, A.: Corrosion and passivation of magnesium alloys. Corrosion Sci. http://dx.doi.org/10.1016/j.corsci.2016.05.041
2. Pardo, A., Merino, M.C., Coy, A.E., Arrabal, R., Viejo, F., Matykina, E.: Corrosion behaviour of magnesium/ aluminium alloys in 3.5 wt% NaCl. Corrosion Sci. **50**, 823–834 (2008)
3. Agnew, S.R.: Wrought magnesium: a 21st century outlook. JOM **56**(5), 20–1 2004
4. Makar, G.L., Kruger, J.: Corrosion of magnesium. Int. Mater. Rev. **38**(3) (1993)
5. Pardo, A., Merino, M.C., Coy, A.E., Viejo, F., Arrabal, R.: Jr SF. Influence of microstructure and composition on the corrosion behaviour of Mg/Al alloys in chloride media. Electrochimica Acta **53**, 7890–902 (2008)
6. Zhao, M., Liu, M., Song, G., Atrens, A.: Influence of pH and chloride ion concentration on the corrosion of Mg alloy ZE41. Corrosion Sci. **50**(11), 3168–78 (2008)
7. Obayi, C.S., Tolouei, R., Mostavan, A., Paternoster, C., Turgeon, S., Okorie, B.A., Obikwelu, D.O., Mantovani, D.: Effect of grain sizes on mechanical properties and biodegradation behaviour of pure iron for cardiovascular stent application. Biomatter. **6**(1) e959874, 9 (2016)
8. Florianovich, G.M.: Electro less dissolution of metals: substantiation and alternative notions. Russ. J. Electrochem. **36**(10), 1037–1042 (2000)

9. Cao,F.-H., Len, V.-H., Zhang, Z., Zhang, J.-Q.: Corrosion behavior of magnesium and its alloy in NaCl solution, Russian J. Electrochem. **43**(7) 837–843 (2007). ISSN 1023-1935. © Pleiades Publishing, Ltd
10. Merino, M.C., Pardo, A., Arrabal, R., Merino, S., Casajús, P., Mohedano, M.: Influence of chloride ion concentration and temperature on the corrosion of Mg–Al alloys in salt fog. Corros. Sci. **52**, 1696–1704 (2010)
11. Pardo, A., Feliu S., Merino, M.C., Arrabal, R., Matykina, E.: Int. J. Corrosion. **2010**(953850) 8 (2010). https://doi.org/10.1155/2010/953850
12. Jiang, J., Zhou, Q., Jisen, Y., Ma, A., Song, D., Fumin, L., Yang, D., Chen, J., Zhang, L.: Comparative analysis for corrosion resistance of micro-arc oxidation coatings coarse-grained and ultra-fine grained AZ91D Mg alloy. Surf. Coat. Technol. **216**, 259–266 (2013)

Chapter 7
Removal of Indoor Air Pollutants Using Activated Carbon—A Review

Prashant P. Bhave and Deepali Yeleswarapu

Abstract Indoor air pollution and its control are a major health concern as a person spends 70% of his day indoors. Adsorption is one of the most preferred air purification systems as it has no by-products. Zeolites, porous clay ore, activated alumina, silica gel, and molecular sieve adsorbents have also been used for removal of the indoor air pollutants. Activated carbon is preferred due to its high microporous volume and adsorption kinetics. Researchers have studied the removal efficiency of activated carbon using various methods under various conditions. Carbon activated chemically or thermally can be used for removal of indoor air pollutants in the form of granules or fibers. The present study compares the effects of activation process, temperature, and relative humidity on the removal efficiency of activated carbon as an indoor air purifier.

Keywords Indoor air · VOC · Adsorption · Activated carbon · Carbon fibers

7.1 Introduction

Indoor environments have a major role to play in the health of an individual, as about 70–90% [1–4] of the day is spent indoors. The indoor air quality is affected by various sources such as air fresheners [5], furniture, building material [6], and activities such as cooking with unconventional fuel, smoking, to name a few. Outdoor air carrying pollutant gases and particulate matter (PM) entering indoor is another source contributing to indoor air pollution [7]. Particulate matter, carbon monoxide, nitrogen dioxide, ozone, and a number of volatile organic compounds (VOCs) such as alkanes, benzene, formaldehyde, toluene, polycyclic aromatic hydrocarbons, terpenes, and phthalates [8, 9] can be found in the indoor environment emitted by a number of consumer products and occupant activities [4]. Most of the VOCs are present in higher concentrations in the indoor air as compared to the outdoors as the infiltrating air from the outside adds to the concentrations present in the indoor environment [6,

P. P. Bhave (✉) · D. Yeleswarapu
Civil & Environmental Engineering Department, VJTI, Mumbai 400019, India
e-mail: ppbhave@ci.vjti.ac.in

© Springer Nature Singapore Pte Ltd. 2020

V. Sivasubramanian and S. Subramanian (eds.), *Global Challenges in Energy and Environment*, Lecture Notes on Multidisciplinary Industrial Engineering, https://doi.org/10.1007/978-981-13-9213-9_7

9]. These pollutants further interact with each other and pose the risk of forming new secondary pollutants [10]. The presence of indoor air pollutants (IAPs) gives rise to symptoms such as drowsiness and sore throat, among others, which are collectively known as sick building syndrome; other than this, some of the VOCs can cause cancer [4, 6, 11]. The effect of IAPs can be reduced by using air purifiers, controlling the air flow rate, and reducing the emission rate, all of which do not give a satisfactory result on their own [12, 13]. Adsorption is one of the most preferred air purification methods as it has no by-products, low energy requirement, and ease of applicability over a relatively wide range of temperatures and pressures [14]. It is also considered to be highly efficient and inexpensive [15]. Adsorption is considered effective even at low concentration levels, i.e., parts per million (ppm) [16]. Adsorption can be carried by using various adsorbent materials such as zeolites, porous clay ore, activated alumina, silica gel, and/or molecular sieve for removal of indoor air pollutants; however, activated carbon is preferred due to its high microporous volume, high degree of surface reactivity and adsorption kinetics and regeneration rates [14, 17]. Activated carbon (AC) is considered to be a useful method for the purification of indoor air as it can adsorb pollutants from gases and vapors in low concentration [18, 19]. A range of factors affects the performance of these AC filters which will be discussed in this paper.

7.2 Activated Carbon Synthesis

Activated carbon can be synthesized by making use of materials that have high carbonaceous content. This material is subjected to carbonization and activation to achieve the final product of activated carbon. Carbonization is the process of eliminating the non-carbon content, whereas activation involves the development of pores in the material [20]. The process of developing activated carbon can be carried out in different methods, namely the physical (thermal) method, chemical method, and steam pyrolysis [21]. The physical method carries out carbonization and activation in two different steps. The chemical method on the other hand deals with both the aspects in one step. This is done by using an activating agent which carbonizes and brings about an increase in the pore volume by dehydration and decomposition [20]. Various chemicals can be used as an activating agent, of which the most common are H_3PO_4, KOH, $ZnCl_2$, and NaOH. The activation leads to formation of AC with high porosity and increases the surfaces area [20, 22, 23]. The adsorption of the organic or pollutant molecules depends on the pore size and pore distribution of the AC. Pores are classified as micropores (size < 2 nm), macropores (>50 nm), and transitional or mesopores (2–50 nm). AC with higher number of micropores and smaller mesopores is suitable for gas phase [24]. Commercially available AC is nonpolar in nature; however, in the case of VOCs, AC's nonpolar nature and hydrophobic nature would show very little removal of hydrophilic pollutants [15]. For the purpose of purification of indoor air, the forms of AC mainly used are granular activated carbon and activated carbon fibers as cloths, felts, or nonwoven fibers.

7.2.1 Granular Activated Carbon (GAC)

Granular activated carbons and powdered activated carbon are both derived from crushed carbon particles. Activated carbon present in the form of granules ranging from the sizes of 0.2–5 mm is classified as GAC, while powder activated carbon essentially ranges in the size of 15–25 micron [25]. GAC, however, is the preferred choice out of the two for gas-phase adsorptions as they have large surface area and a highly developed pore structure [26, 27]. GAC is prepared using coconut shells, wood, or coal as precursors; however, waste agricultural products such as corn cobs, groundnut shells, maize cobs, and sugarcane bagasse along with other substances such as sludge, sawdust, coir pith, and fly ash can be used as well.

7.2.2 Activated Carbon Fibers

Activated carbon fibers are better than the GACs in a number of ways as they have higher pore volume, largely containing micropores, larger surface area, and a more uniform distribution of pore structure, faster kinetics than GAC [11, 28, 29]. Other than these properties, regeneration of ACFs can be easily carried by electrothermal regeneration. Activated carbon fiber can be synthesized using some specific textile fabrics or various others precursors. Precursors used for ACF are phenolic resins, polyamides, polyacrylamides, pitch, cellulose-based fibers, and viscose rayon fibers, to name a few [30]. Fibers can be made available in the form of either felts or cloths depending on whether they have been woven or not. The pore structure gets affected depending on the nature of weave as well [31].

7.3 Factors Affecting Adsorption

7.3.1 Surface Structure

Adsorption depends on the surface microstructure of the activated carbon [18]. For gaseous-phase adsorption, there is a requirement of high pore volumes. However, for adsorption of gases at low concentrations, preferential filling of micropores takes place; hence, AC with high micropore volume gives better results; especially, for VOC molecules, narrow micropores are required, and the saturation of same affects the rate of removal [12, 32, 33]. Supporting this are the findings of a study, where it was noted that benzene fills narrow micropores of size less than 0.7 nm, whereas toluene occupies micropores of different sizes [34]. A study was conducted by C. Lorimier et al., consisting of felts having an inter-fiber mean diameter of 65 μm and cloths having a mean diameter of 121 μm between the yarns and a diameter of 13 μm for pores between the yarns. It was found that toluene does not fill pores smaller than its molecular size, thus implying that a greater hidden micropore volume does not necessarily mean greater adsorption. At low adsorbate concentration, preferential filling of smaller micropores takes place resulting in higher adsorption of VOC on the felts at a higher concentration but greater adsorption on the cloths at lower concentration (80 mg/cu m). This phenomenon of preferential filling is believed to occur due to the overlap of attractive forces of opposite pore walls [11, 12]. Micropore sizes are generally uniform in ACFs and in any case are dependent on the activation methods and agents employed [30]. Thermal activation brings about burnoff in the AC precursor which affects the pore structure. The rate of burnoff depends on the temperature which in turn affects the pore structure, with higher burnoffs leading to formation of wider pores [30]. In case of thermal activation of AC, activation under CO_2 flow rather than steam gives rise to higher microporosity; however, the latter saves time [35]. At higher temperatures, aromatization process takes place which results in enrichment of carbon content [29]. On viscose rayon fabric, impregnation with $ZnCl_2$ and H_3PO_4 gives high porosity, with the latter forming wider micropores as compared to the former [36]. AC can be modified by using various acids and alkalis post-activation. Treatment with acids gives rise to increase in the pore size but reduces the overall volume of pores, whereas alkalis increase the pore volume and leave the mean pore size somewhat unchanged [19]. Etching is another form of surface modification which can be brought about by the use of ammonia on AC. Etching causes decomposition of surface oxygen leading to vacant sites on AC, thus bringing about surface modification [37].

7.3.2 Surface Chemistry

Depending on the precursor, activating procedure, and agents, the surface chemistry varies. AC, however, is desired to be nonpolar and hydrophobic in nature [15]. The presence of carbonyls showed an improvement in toluene removal efficiency [32], whereas the presence of oxygen groups tends to decrease the removal efficiency for gases like o-xylene and benzene [19, 26]. In another study evaluating the uptake of benzene, toluene, ethylbenzene, and o-xylene (BTEX), the results show the uptake efficiency being affected by surface groups such as carboxylates and their hydration and the presence of acidic surface oxides [38]. Adsorption of methyl ethyl ketone (MEK) and methanol is higher than that of benzene as a result of chemisorption due to the presence of certain groups, whereas benzene exhibited only diffusion [39]. Activation with acids increases the surface oxides in terms of phenolic, carboxylic, and carbonyl groups, whereas alkalis reduce the number of oxygen groups formed on the surface [32]. The groups that are formed on the surface can be identified by means of absorbance of infrared light. These groups can, however, be modified by means of thermal or chemical treatment. In order to increase the surface oxygen groups, the AC must be treated with acids, whereas treatment with alkalis results in the reduction of the same [19]. Surface oxidation can also be brought about by subjecting the AC to the gaseous stream of O_2, or with the help of chemical solutions of ammonia, urea, or any nitrogen-containing solution [40]. Activation with ammonia, however, can lead to the formation of amides and aromatic amides or pyridines and protonated pyridines depending on the temperature of activation [37]. The use of cupric oxide (CuO) impregnated on ACF helps in abating the effect of high relative humidity (RH). Removal efficiency of gases across a wide range on RH (30–70%) was found to be similar [41]. It must be noted that change in the chemical composition may result in the modification of pore structure as well [42].

7.3.3 Temperature

An increase in temperature showed a decrease of 22–26% in the adsorption capacity of toluene at 303 K (30 °C) as compared to that at 298 K (25 °C) [43]. Breakthrough curves were found to be steeper and occur early at higher temperature (363 K) partly due to reaction rate constants [44]. Similar results were found for MEK and benzene adsorption. The breakthrough time fell from close to 250 min to a little over 100 min as the temperature was raised from 30 to 60 °C [45]. As the temperature increases, the adsorbed gas decreases at a given relative pressure. However, sometimes, the amount of adsorbed gas is higher at a given temperature than at a lower temperature; this effect is termed as activated entry effect. This temperature dependency is speculated to be caused by the quick approach to equilibrium at high temperatures, beyond the equilibrium; the elevated temperature would not affect the adsorption [18]. Higher temperature also assists in diffusion of gas, thus allowing it an entry into micropores' narrow constrictions [29]. It was found that at low temperatures (278–303 K), gas undergoes physical adsorption; thus, the chemical nature does not play any significant role; however, this does not hold true for higher temperatures (318–353 K) [18].

7.3.4 Relative Humidity

Water vapor is present in the indoor environment and is adsorbed onto AC and alters the adsorption capacity either by competitive co-adsorption or cooperative co-adsorption [17, 46]. Adsorption of hydrophobic VOCs such as benzene and toluene which fill in the micropores of AC gets affected due to the presence of RH. Water molecules get adsorbed on AC by means of reaction with the surface oxides present on AC, forming a layer and progressing toward the formation of clusters. These clusters fill the micropores and sometimes block them causing a hindrance in the adsorption of nonpolar compounds [26, 39, 47]. The blocking of the micropores takes place due to capillary condensation which reduces the diffusion of the gases [39, 48]. It has been observed that as the RH increases, the water vapor molecules start to fill in larger-sized mesopores, other than the micropores that have been occupied by the same [48]. In a study conducted by Gallego, it was found that the average reduction of VOCs adsorbed in the presence of RH variation from 26 to 46% showed a significant decrease in the removal efficiency [17]. The presence of RH in the indoor environment affects the removal efficiency of hydrophobic VOCs, usually close to or above 50% as compared to relatively dry air with RH at 20% or less [27]. In another study involving adsorption of NO and BTEX, it was found that as the humidity increased from 2100 to 22,000 ppmv (i.e., parts per million volume), the removal efficiency dropped from above 25% to less than 10% for all the gases tested [49]. The case for hydrophilic compounds is different as reported for acetone, where removal fell from 100 ppm at 0 RH to about 80 ppm at RH = 90%. In the same study, it was observed that the removal of benzene fell from about 340 ppmv (RH = 0) to a little above 200 ppmv (RH = 80%), showing that the presence of high RH affects hydrophobic compounds adversely as compared to the hydrophilic ones [46]. However, methanol which is a polar compound occupies the vacant sites by chemisorption available and on saturation of those sites occupies the remaining by means of dispersion leading to a drastic fall in the adsorption capacity of AC [39]. Due to the nature of competitive adsorption and occupation of available micropore sites, water vapor reduces the breakthrough time for AC and could lead to frequent regeneration or replacement [50].

7.3.5 Ozone

In the presence of ozone, various VOCs form secondary products such as limonene aldehyde, ketolimonon (limonene, which is a common VOC found in indoor environment) were found as a result of oxidation. These products as a result of competitive adsorption or desorption may be reintroduced in the air stream, especially in the case of filters that recirculate air [51]. Reaction of ozone with the AC could lead to the formation of surface oxides or gasification, wherein the carbon content is lost as it gets oxidized to carbon dioxide [47, 52]. AC with ozone adsorbed shows affinity toward polar compounds, thus increasing the adsorption of water vapor on the surface of AC [47, 51]. In a study to evaluate the effect of ozone on the adsorption of limonene on AC, it was found that about 42% and 13% were found to remain on the AC, suggesting that the rest was oxidized to form multiple secondary products. However, it was added that some of the products formed could be due to impurities present on the AC [51]. In another study, breakthrough curves for toluene adsorption were plotted for three different cases which were adsorption in the absence of ozone, co-adsorption with ozone, and adsorption on AC preloaded with ozone. It was found that ozone reduces the breakthrough time when preloaded on to the AC, whereas in the case of co-adsorption, initially it shows a trend similar to that of the ozone free adsorption; however, owing to the increase in ozone concentration on AC as the adsorption progresses, it reduces and has a much shorter breakthrough time [47].

7.4 Antimicrobial Application of Activated Carbon

Water vapor present in the air gets adsorbed on the AC along with various gases. The moist conditions provide a favorable environment for the growth of microbes [53]. Even though ACs do not adsorb microorganisms, they get deposited with the stream of gases entering the filter [3]. In case of re-circulation of air, there are chances of microbes being suspended into the air. Furthermore, the organic and inorganic gases that get adsorbed on the AC act as a substrate and enable multiplication of bacteria [54]. The use of silver (Ag) particles is considered an effective method for disinfection by means of impregnation on AC [54]. As the particle size of Ag reduces to nanometer scale, it is believed to develop high surface area and hence becomes highly reactive [55] Impregnation of Ag on AC using the ion-dipole method shows a 99% removal efficiency of bacteria within a contact time of 120 min; however, a fall in adsorption capacity of VOCs is a downside. It was, however, observed that when the impregnated Ag particles were kept at a concentration of 0.98% wt or lesser, the VOC adsorption efficiency was not adversely affected [3].

7.5 Activated Carbon Regeneration

In order to prolong the life of AC, regeneration is practiced where the gases collected on it are desorbed by one of the following methods such as thermal desorption or by making use of ozone and microwave heating or electrochemical desorption. Every time regeneration is carried out, some amount of the adsorption capacity of AC is lost since all the gases adsorbed on it are not removed, and some amount of AC gets burned and needs to be disposed [56, 57]. Activated carbon cloths are electrically conductive and can be regenerated using electrical current which in turn heats the cloth leading to desorption, which could lead to regeneration to the extent of above 95% [29]. Regeneration of the AC depends on the nature of gas adsorbed. Regeneration at 150 °C shows a fall in the adsorption capacity of formaldehyde, an almost consistent efficiency across consecutive cycles for limonene, undecane, and xylene, whereas the fall in flow rate affected the removal efficiency of benzene and toluene negatively in consecutive cycles [56]. Regeneration can also be carried out by using outdoor air streams which usually have lower VOC concentrations. However, this could lead to decrease in removal efficiency in subsequent cycles as found for VOCs such as benzene, toluene, o-xylene, 1-butanol, undecane, and formaldehyde, where the removal efficiency fell from an initial percentage of 40–70% to 25–40% in the subsequent cycles [56]. Thermal regeneration at 450 °C of AC led to a loss of 7.2, 13.8, and 20.3% in carbon content from triethylamine, *n*-decane, and limonene saturated ACs, respectively. A loss in micropore volume was also recorded at 50, 36, and 58%, respectively, for the above-mentioned VOCs [57]. The use of catalysts, cerium oxide along with titanium dioxide (CeO/TiO_2) along with AC where TiO_2 acts as a binder and CeO as the catalyst, can be used for adsorption of VOCs and then the oxidation of the said gases succeeding desorption from AC at 450 °C. Oxidation shall convert all the gases to CO_2 and H2O [57]. Regeneration of AC can be established by impregnating AC with TiO2, to enable photodegradation of the organic pollutants. Calcination can be employed to ensure adherence of TiO_2 particles on AC [49, 58]. AC concentrates the pollutants around the TiO2 giving rise to increased photodegradation rate. At low RH (around 2100 ppmv), diffusion of pollutants from AC to the TiO_2 sites is easier. However, even in high RH conditions, the adsorption capacity of TiO_2/AC for BTEX remained in the range of 70–75% [49]. At low concentration of 0.5 ppm, the removal of HCHO by AC/TiO_2 showed almost 100% removals. The effect of TiO_2, however, depends on the flow velocity, since it affects the residence time. The same study, however, found that TiO_2 particles dispersed on activated carbon can hinder adsorption on AC due to blocking of pores, thus affecting the adsorption efficiency [58].

7.6 Conclusions

There have been numerous researches conducted on monitoring of real-time emissions and levels of a wide variety of IAPs that arise from various occupant activities such as smoking and cooking [59, 60]. However, studies conducted on AC for the removal of IAPs have been mostly carried out under controlled laboratory environments and more commonly for the most prevalent VOCs, namely limonene, toluene, and benzene. A study conducted on removal of IAPs emitted due to the occupant activities could give a fair idea about the use of AC as an indoor air purifier for practical purposes. Based on the present study on removal of indoor air pollutants using activated carbon, the following are the conclusions drawn upon:

Properties of activated carbon depend on the activation method used, and the activation temperature and activating agent play a crucial role that determines the efficiency of the AC.
Improvement in the efficiencies of activated carbon used for removal of indoor air pollutants can be brought about by employing surface modification to bring about desirable changes in porosity and surface chemistry of AC.
While GAC can be modified to better suit the criteria of the high microporous area, activated carbon fibers have uniform pore structure and are preferred over the former.
Ozone is one of the important gases present in the indoor environment which affects the performance of AC.
Relative humidity plays a very crucial role in the removal efficiency of the organic compounds and the life of AC, with higher humidity levels causing hindrance to the performance. This, however, can be tackled by impregnation of TiO_2.
The presence of humidity could lead to microbial growth on the AC which can be removed by making use of Ag or Ag nanoparticles.

In situ regeneration of AC can be carried out to improve the life of AC by various methods; however, regeneration affects the surface structure and efficiency as desorption of gases does not take place completely.

References

1. Spengler, J.D., Sexton, K.: Indoor air pollution: a public health perspective. Am. Assoc. Adv. Sci. **221**(4605), 9–17 (1983)
2. Menichini, E., Iacovella, N., Monfredini, F., Turrio-Baldassarri, L.: Relationships between indoor and outdoor air pollution by carcinogenic PAHs and PCBs. Atmos. Environ. **41**(40) 9518–9529 (2007)
3. Pei, L., Zhou, J., Zhang, L.: Preparation and properties of Ag-coated activated carbon nanocomposites for indoor air quality control. Build. Environ. **63**, 108–113 (2013)
4. Wang, S., Ang, H.M., Tade, M.O.: Volatile organic compounds in indoor environment and photocatalytic oxidation: state of the art. Environ. Int. **33**(5), 694–705 (2007)
5. Steinemann, A.: Ten questions concerning air fresheners and indoor built environments. Build. Environ. **111**, 279–284 (2017)
6. Missia, D.A., Demetriou, E., Michael, N., Tolis, E.I., Bartzis, J.G.: Indoor exposure from building materials: a field study. Atmos. Environ. **44**(35), 4388–4395 (2010)
7. Ciuzas, D., et al.: Characterization of indoor aerosol temporal variations for the real-time management of indoor air quality. Atmos. Environ. **118**, 107–117 (2015)
8. Pei, X.Q., Song, M., Guo, M., Mo, F.F., Shen, X.Y.: Concentration and risk assessment of phthalates present in indoor air from newly decorated apartments. Atmos. Environ. **68**, 17–23 (2013)
9. Penney, D., Benignus, V., Kephalopoulos, S., Kotzias, D., Kleinman, M., Verrier, A.: Guidelines for indoor air quality. WHO Guidel. **9** 454 (2010)
10. Singer, B.C., et al.: Indoor secondary pollutants from cleaning product and air freshener use in the presence of ozone. Atmos. Environ. **40**(35), 6696–6710 (2006)
11. Lorimier, C., Subrenat, A., Le Coq, L., Le Cloirec, P.: Adsorption of toluene onto activated carbon fibre cloths and felts: application to indoor air treatment. Environ. Technol. **26**(11) 1217–1230 (2005)
12. Foster, K.L., Fuerman, R.G., Economy, J., Larson, S.M., Rood, M.J.: Adsorption characteristics of trace volatile organic compounds in gas streams onto activated carbon fibers. Chem. Mater. **4**(5), 1068–1073 (1992)
13. Yu, Q.L., Brouwers, H.J.H.: Applied Catalysis B: environmental Indoor air purification using heterogeneous photocatalytic oxidation. Part I: experimental study. Appl. Catal. B Environ. **92**, 454–461 (2009)
14. Shafeeyan, M.S., Mohd, W., Wan, A., Houshmand, A., Shamiri, A.: A review on surface modification of activated carbon for carbon dioxide adsorption. J. Anal. Appl. Pyrolysis. **89**(2) 143–151 (2010)
15. Zhang, X., Gao, B., Creamer, A.E., Cao, C., Li, Y.: Adsorption of VOCs onto engineered carbon materials: A review. J. Hazard. Mater. **338**, 102–123 (2017)
16. Das, D., Gaur, V., Verma, N.: Removal of volatile organic compound by activated carbon fiber. **42** 2949–2962 (2004)
17. Gallego, E., Roca, F.J., Perales, J.F., Guardino, X.: Experimental evaluation of VOC removal efficiency of a coconut shell activated carbon filter for indoor air quality enhancement. Build. Environ. **67**, 14–25 (2013)
18. Chiang, Y.-C., Chiang, P.-C., Huang, C.-P.: Effects of pore structure and temperature on VOC adsorption on activated carbon. Carbon N. Y. **39**(4), 523–534 (2001)

19. Li, L., Liu, S., Liu, J.: Surface modification of coconut shell based activated carbon for the improvement of hydrophobic VOC removal. J. Hazard. Mater. **192**(2) 683–690 (2011)
20. Caturla, F., Molina-Sabio, M., Rodríguez-Reinoso, F.: Preparation of activated carbon by chemical activation with $ZnCl_2$. Carbon N. Y **29**(7) 999–1007 (1991)
21. Ioannidou, O., Zabaniotou, A.: Agricultural residues as precursors for activated carbon production-a review. Renew. Sustain. Energy Rev. **11**(9), 1966–2005 (2007)
22. Adegoke, K.A., Bello, O.S.: Dye sequestration using agricultural wastes as adsorbents. Water Resour. Ind. **12**, 8–24 (2015)
23. Dias, J.M., Alvim-Ferraz, M.C.M., Almeida, M.F., Rivera-Utrilla, J., Sanchez-Polo, M.: Waste materials for activated carbon preparation and its use in aqueous-phase treatment: a review. J. Environ. Manage. **85**(4), 833–846 (2007)
24. Thomas, W.J., Crittenden, B.: Adsorption Technology and Design. Elsevier Science & Technology Books, London, United Kingdom (1998)
25. Aktaş, Ö., Ferhan, Ç.: Fundamentals of Adsorption onto Activated Carbon in Water and Wastewater Treatment. Wiley-VCH Verlag GmbH & Co KGaA, Germany (2012)
26. Bansal, R.C.R.C., Goyal, M., Roop, C.B., Meenakshi, G., Bansal, R.C.R.C., Goyal, M.: Activated Carbon Adsorption. Taylor & Francis Group (2005)
27. Haghighat, F., Lee, C.S., Pant, B., Bolourani, G., Lakdawala, N., Bastani, A.: Evaluation of various activated carbons for air cleaning—towards design of immune and sustainable buildings. Atmos. Environ. **42**(35), 8176–8184 (2008)
28. Lordgooei, M., Carmichael, K.R., Kelly, T.W., Rood, M.J., Larson, S.M.: Activated carbon cloth adsorption-cryogenic system to recover toxic volatile organic compounds. Gas Sep. Purif. **10**(2) 123–130 (1996)
29. Ramos, M.E., Bonelli, P.R., Cukierman, A.L., Ribeiro Carrott, M.M.L., Carrott, P.J.M.: Adsorption of volatile organic compounds onto activated carbon cloths derived from a novel regenerated cellulosic precursor. J. Hazard. Mater. **177**(1–3) 175–182 (2010)
30. Carrott, P., Nabais, J., Ribeiro Carrot, M., Pajares, J.A.: Preparation of activated carbon fibres from acrylic textile fibres. Carbon N. Y. **39**(10) 1543–1555 (2001)
31. Kim, C., Kang, S., Pui, D.Y.H.: Removal of airborne sub-3 nm particles using fibrous filters and granular activated carbons. Carbon N. Y. **104**, 125–132 (2016)
32. Tham, Y.J., Latif, P.A., Abdullah, A.M., Shamala-Devi, A., Taufiq-Yap, Y.H.: Performances of toluene removal by activated carbon derived from durian shell. Bioresour. Technol. **102**(2), 724–728 (2011)
33. Ramos, M.E., Bonelli, P.R., Cukierman, A.L., Carrott, M.M.L.R., Carrott, P.J.M.: Influence of thermal treatment conditions on porosity development and mechanical properties of activated carbon cloths from a novel nanofibre-made fabric. Mater. Chem. Phys. **116**(2–3), 310–314 (2009)
34. Lillo-Ródenas, M.A., Cazorla-Amorós, D., Linares-Solano, A.: Behaviour of activated carbons with different pore size distributions and surface oxygen groups for benzene and toluene adsorption at low concentrations. Carbon N. Y. **43**(8), 1758–1767 (2005)
35. Lillo-Ródenas, M.A., Cazorla-Amorós, D., Linares-Solano, A.: Benzene and toluene adsorption at low concentration on activated carbon fibres. Adsorption **17**(3), 473–481 (2011)
36. Huidobro, A., Pastor, A.C., Rodr guez- Reinoso, F.: Preparation of activated carbon cloth from viscous rayon. Carbon N. Y. **39**(3) 389–398 (2001)
37. Mangun, C.L., Benak, K.R., Economy, J., Foster, K.L.: Surface chemistry, pore sizes and adsorption properties of activated carbon fibers and precursors treated with ammonia. Carbon N. Y. **39**, 1809–1820 (2001)
38. Daifullah, A.A., Girgis, B.: Impact of surface characteristics of activated carbon on adsorption of BTEX. Colloids Surfaces A Physicochem. Eng. Asp. **214**(1–3), 181–193 (2003)
39. Rodríguez-Mirasol, J., Bedia, J., Cordero, T., Rodríguez, J.J.: Influence of water vapor on the adsorption of VOCs on Lignin-based activated carbons. Sep. Sci. Technol. **40**(15), 3113–3135 (2005)
40. Stavropoulos, G.G., Samaras, P., Sakellaropoulos, G.P.: Effect of activated carbons modification on porosity, surface structure and phenol adsorption. J. Hazard. Mater. **151**(2–3), 414–421 (2008)

41. Huang, Y.C., Luo, C.H., Yang, S., Lin, Y.C., Chuang, C.Y.: Improved removal of indoor volatile organic compounds by activated carbon fiber filters calcined with copper oxide catalyst. Clean Soil Air Water **38**(11), 993–997 (2010)
42. Shim, J.W., Park, S.J., Ryu, S.K.: Effect of modification with HNO_3 and NaOH on metal adsorption by pitch-based activated carbon fibers. Carbon N. Y. **39**(11), 1635–1642 (2001)
43. Carratalá-Abril, J., Lillo-Ródenas, M.A., Linares-Solano, A., Cazorla-Amorós, D.: Activated carbons for the removal of low- concentration gaseous toluene at the semipilot scale. Ind. Eng. Chem. Res. **48**(4), 2066–2075 (2009)
44. Chuang, C.L., Chiang, P.C., Chang, E.E.: Modeling VOCs adsorption onto activated carbon. Chemosphere **53**(1), 17–27 (2003)
45. Huang, Z.H., Kang, F., Liang, K.M., Hao, J.: Breakthrough of methyethylketone and benzene vapors in activated carbon fiber beds. J. Hazard. Mater. **98**(1–3), 107–115 (2003)
46. Cal, M.P., Rood, M.J., Larson, S.M.: Gas phase adsorption of volatile organic compounds and water vapor on activated carbon cloth. Energy Fuels **11**(2), 311–315 (1997)
47. Kingsley, M.L., Davidson, J.H.: Adsorption of toluene onto activated carbons exposed to 100 ppb ozone. Carbon N. Y. **44**(3), 560–564 (2006)
48. Pei, J., Zhang, J.S.: Determination of adsorption isotherm and diffusion coefficient of toluene on activated carbon at low concentrations. Build. Environ. **48**(1), 66–76 (2012)
49. Ao, C.H., Lee, S.C.: Enhancement effect of TiO_2 immobilized on activated carbon filter for the photodegradation of pollutants at typical indoor air level. Appl. Catal. B Environ. **44**(3), 191–205 (2003)
50. Li, J., Li, Z., Liu, B., Xia, Q., Xi, H.: Effect of relative humidity on adsorption of formaldehyde on modified activated carbons. Chinese J. Chem. Eng. **16**(6), 871–875 (2008)
51. Metts, T.A.: Heterogeneous reactions of ozone and d-limonene on activated carbon. Indoor Air **17**(5), 362–371 (2007)
52. Metts, T.A., Batterman, S.A.: Effect of VOC loading on the ozone removal efficiency of activated carbon filters. Chemosphere **62**(1), 34–44 (2006)
53. Khajavi, R., Bahadoran, M.M.S., Bahador, A., Khosravi, A.: Removal of microbes and air pollutants passing through nonwoven polypropylene filters by activated carbon and nanosilver colloidal layers. J. Ind. Text. **42**(3), 219–230 (2013)
54. Yoon, K.Y., Byeon, J.H., Park, C.W., Hwang, J.: Antimicrobial effect of silver particles on bacterial contamination of activated carbon fibers. Environ. Sci. Technol. **42**(4), 1251–1255 (2008)
55. Tuan, T.Q., et al.: Preparation and properties of silver nanoparticles loaded in activated carbon for biological and environmental applications. J. Hazard. Mater. **192**(3), 1321–1329 (2011)
56. Sidheswaran, M.A., Destaillats, H., Sullivan, D.P., Cohn, S., Fisk, W.J.: Energy efficient indoor VOC air cleaning with activated carbon fiber (ACF) filters. Build. Environ. **47**(1), 357–367 (2012)
57. Raso, R.A., Zeltner, M., Stark, W.J.: Indoor air purification using activated carbon adsorbers: regeneration using catalytic combustion of intermediately stored VOC. Ind. Eng. Chem. Res. **53**(9), 3622–3628 (2014)
58. Lu, Y., Wang, D., Ma, C., Yang, H.: The effect of activated carbon adsorption on the photocatalytic removal of formaldehyde. Build. Environ. **45**(3), 615–621 (2010)
59. Höllbacher, E., Ters, T., Rieder-Gradinger, C., Srebotnik, E.: Emissions of indoor air pollutants from six user scenarios in a model room. Atmos. Environ. **150**, 389–394 (2017)
60. Uhde, E., Schulz, N.: Impact of room fragrance products on indoor air quality. Atmos. Environ. **106**, 492–502 (2015)

Chapter 8
Mercury Air Pollution from Coal Combustion in India and Its Control Measures

Prashant P. Bhave and Rajita Shrestha

Abstract Mercury exists in the earth's crust mostly as cinnabar (HgS) and hence is present as an impurity in most of the fuel and minerals. Coal is the primary fuel source for 40.7% of the world according to 2014 data. In recent years, there is a decline in coal usage in countries like Europe, USA, and China. However, India's energy consumption has doubled since year 2000 with coal being the major fuel source. With the rising coal demand, India is predicted to be the largest coal importer in the world before 2020. Mercury content in coal varies from 0.05 to 0.2 g/MT according to the United Nations Environmental Program's mercury estimation toolkit. Mercury pollution monitoring and its intercontinental transport modeling done for Arctic region concludes that 32% of this pollution originated from the Asian countries. Various world mercury emission estimates done in the past researches consider India to be the second largest mercury emitter after China. Last mercury emission estimate for India was done in the year 2004. This paper compares the mercury emissions from coal combustion in India to that of the world and studies the effectiveness of various air pollution control devices in the removal of mercury from coal combustion emissions.

Keywords Mercury · Coal combustion · Long-range air pollution · Emission estimate · Air pollution control device

8.1 Introduction

Mercury exists in the environment in three forms elementary mercury vapors, ionic mercury, and organic methyl mercury [1]. Most of the mercury in nature is elemental Hg vapor which circulates in the atmosphere for up to one year and hence can be widely dispersed around the globe. Inorganic mercury which is emitted into the atmosphere associated with airborne particles or in gaseous form is removed by dry or wet deposition over short distances (few tens to few hundreds of kilometer) [2]. Inorganic

P. P. Bhave (✉) · R. Shrestha
Civil and Environmental Engineering Department, VJTI, Mumbai 400019, India
e-mail: ppbhave@vjti.org.in

© Springer Nature Singapore Pte Ltd. 2020
V. Sivasubramanian and S. Subramanian (eds.), *Global Challenges in Energy and Environment*, Lecture Notes on Multidisciplinary Industrial Engineering, https://doi.org/10.1007/978-981-13-9213-9_8

mercury deposited changes its form to organic methylmercury by various naturally occurring biological processes [3]. This methyl mercury can bio-concentrate more than a million-fold in the aquatic food chain, and hence, mercury emissions become a major health concern in the environment [2]. The transport and deposition of mercury from the source depend on the chemical form of mercury emitted, the stack height, the area surrounding the site, topography, and meteorology [2, 4]. Minamata Convention, which was established in October 2013, has an objective to protect human health and environment from anthropogenic emissions and releases of mercury and mercury compounds by a set of control and phase-out regulations [5]. India joined the convention on September 30, 2014. However, India lacks a reliable data source for mercury used in industrial processes and products. Elemental mercury vapors enter the bloodstream on inhalation while inorganic mercury enters the bloodstream through the gastrointestinal system and dermal contact [1]. Elemental mercury is highly lipophilic which allows it to cross blood–brain and placental barriers [1]. Inorganic mercury exposure damages kidney [4]. Organic methylmercury exposure to humans occurs through ingestion of polluted food items, especially fish. Critical target for methylmercury toxicity is the nervous system. A fetus, newborn, and young children are at a higher risk of developing neurological diseases due to methylmercury poisoning [4]. Individuals with past history of diseases related to liver, kidneys, and lungs are also at higher risk [4].

8.2 Intercontinental Mercury Transportation

Intercontinental transport of mercury is estimated using modeling studies. These models require various measurements of atmospheric mercury concentrations, airborne particulate mercury concentration, and concentrations of mercury deposited through wet deposition [6]. The modeling also requires data like wind direction, mixing height, various mercury transformations in the atmosphere, dry deposition velocity of mercury forms, and mercury emission estimate data [7]. Global anthropogenic atmospheric mercury modeling studies conducted in the year 2014 using GEOS-Chem global chemical transport model estimated East Asia to be the primary source of mercury pollution for all the continents. This study also evaluated Indian subcontinent to be the second largest contributor to particulate mercury around the globe [8]. Similar study conducted using emission data of 1996 concluded that 67% of total depositions to North America originated from outside the continent including 24% from Asin sources. About half the mercury deposition in Arctic regions originated from other continents majorly Asia and Europe [9]. Anthropogenic mercury emissions to atmosphere and releases to environment can be associated with various mercury sources. Various mercury sources are divided into broad categories which includes extraction and combustion of fuels, primary metal production, production of other materials with mercury impurities (cement, pulp and paper, and lightweight aggregates), intentional use of mercury in industrial processes (chlor-alkali, vinyl chloride monomer), intentional use in consumer products (batteries, thermometers, paints, and biocides), and other intentional uses like dental amalgam fillings and production of recycled metals [10]. Various studies in the past have estimated global mercury emissions from various sources over the years. In 1995, a worldwide mercury emissions estimate concluded that 3/4th of the total emissions (56%) were caused due to fossil fuel combustion in China, India, and Korea [11]. Estimates of year 2005 states that out of a total 1930 tonnes of mercury released from anthropogenic sources, 400 tonnes comes from fossil fuel combustion in China and 180 tonnes come from fossil fuel combustion in India [12]. Such studies in the past have recognized coal combustion to be a major source of both elemental and oxidized mercury in the atmosphere. Mercury emissions estimation in India for the period 2000–2004 quantifies mercury release from coal-fired power plants and residential and commercial boilers to be 104.09 and 124.55 tonnes, respectively [13].

8.3 Mercury Emissions from Coal in India and World

Coal occurring in India is of Gondwana origin, low in hydrogen, and high in nitrogen content. Most of them are characterized by high ash content, low Sulfur content, and low chloride content. Most of the coal reserves are confined to eastern and south-central parts of the country in states of Jharkhand, Orissa, Chhattisgarh, West Bengal, Andhra Pradesh, Maharashtra, and Madhya Pradesh [14]. India's coal consumption in the period 2005–2012 increased from 6% of world production to 7.6% of world production in 2012 (Table 8.1) [15]. Within India, consumption of coal is distributed in sectors like steel and washery, power generation, cement, textile, fertilizer and chemicals, paper, brick, colliery consumption, and others. Among the various sectors, power generation consumes the largest fraction of coal available for

Table 8.1 Coal production in the world and coal and lignite consumption in India (million MT)

Year	2005	2006	2007	2008	2009	2010	2011	2012
World	7401.83	7838.974	8226.262	8359.865	8408.690	8920.713	9482.176	9731.362
India (Coal)	442.536	430.832	456.397	550.318	597.64	576.03	637.017	692.441
India (Lignite)	30.087	31.285	33.980	32.421	34.071	37.740	43.105	46.598
Raw coal used in PP	316.716	331.578	360.154	391.692	406.39	382.664	403.905	480.81
Lignite used in PP	23.074	23.92	21.665	25.332	27.92	0	33.09	37.308

Reference Coal Controller's Organisation, GOI, Ministry of Coal, Provisional Coal Statistics

consumption in the country. In 2005, 71.56% of the available raw coal was consumed in power generation which increased to 78.91% in 2007 after which it decreased to 71.17% in 2008 and reached 63.4% in the year 2011. After 2012, coal consumption in power generation increased again to 69.43%. Lignite consumption by the power generation industry also averaged at around 65% of total lignite available [16–23]. Mercury emissions from coal combustion estimation require data regarding activity rate of coal, emission rate, and output factor or distribution factor [10]. Activity rate for coal combustion is derived from the national provisional coal statistics which is published every year. Input factor in this study is decided as 0.005–0.5 g/MT (0.376 g/MT) which has been decided referring the UNEP toolkit for identification and quantification of mercury releases as well as on Mukherjee report on mercury emissions from industrial sources in India and its effects on the environment. Selection of an output factor is done based on the abatement measures for various pollutants included in the coal combustion facilities. Almost all thermal power plants in India are equipped with electrostatic precipitators with more than 99% efficiency [24]. Output factor for this study has been taken from UNEP toolkit for estimation of mercury emissions from anthropogenic sources. Output factor for mercury emissions to air from coal combustion is taken as 0.75 in case of coal used in power plant (PP) combustion and 0.9 for rest of the activities.

Mercury emission graph shows an increasing trend of mercury emissions from India as compared to a decreasing trend in world mercury emissions (Table 8.2). Considering median values for both global and Indian emission data, India contributed to 15% of world mercury emissions in 2005 which increased to 16.02% in 2006 and 37.25% in 2010. This trend can be associated with increasing coal consumption in India (Table 8.2). India's coal consumption rate is estimated to increase further in coming years which implies the need to include effective mercury emissions control devices in our power plants (PP) and industrial boilers (Tables 8.1 and 8.2).

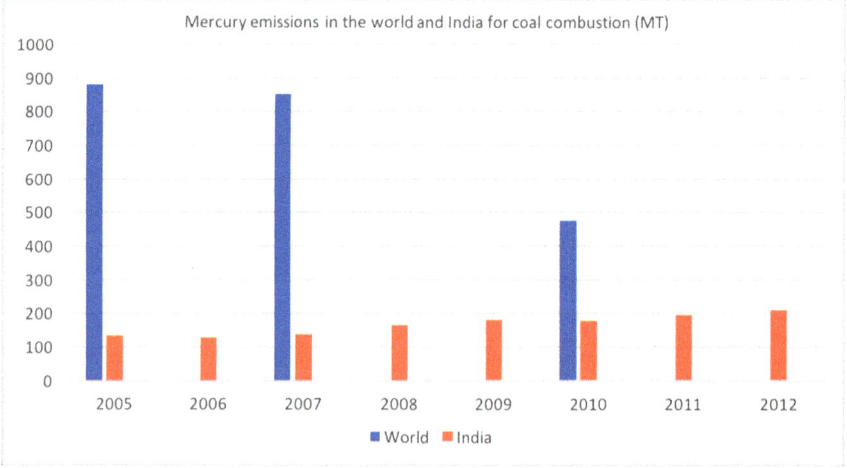

Table 8.2 Mercury emissions over the years in the world and India from coal combustion (MT)

	Year	2005	2006	2007	2008	2009	2010	2011	2012
World		880.2		85.4			474.337		
India (Coal) from PP	Lower limit	12.735	13.327	14.31	15.62	16.277	14.34	16.38	19.42
	Median limit	91.04	95.29	103.18	112.34	116.69	107.911	116.38	138.37
	Upper limit	122.22	127.93	138.29	150.67	156.57	143.49	156.42	185.89
India (other industrial sources)	Lower limit	5.977	4.797	4.884	7.457	8.882	10.391	10.94	9.93
	Median limit	43.2	34.25	33.66	54.308	65.269	68.82	79.781	72.44
	Upper limit	57.87	45.98	45.52	72.656	87.167	93.802	106.702	96.90
Total	Median limit	134.24	129.54	136.84	166.648	181.959	176.731	196.161	210.81

8.4 Mercury Emissions Control Measures

Statistics regarding mercury emissions in the above data show that majority of the mercury in the atmosphere is released due to coal combustion in power plants (PP). A study on mercury emissions from three power plants in India, two public sectors, and one private sector was carried out each having boiler with unit generation capacities of 210, 250, and 500 MW. These plants are equipped with electrostatic precipitator for particulate capture with outlet of electrostatic precipitator (ESP) having a flue gas temperature of 150, 130, and 127 °C. Results for these plants suggested that 60–70% of mercury emissions in flue gases are in the form of gaseous elemental mercury. This is due to low chloride content in the feed coal which does not oxidize mercury vapors [24]. Mercury emissions' reduction from coal combustion in power plants depends on the mercury content of the coal feed, species of mercury emitted, efficiency of the boiler, and type of pollution control measures employed in the plant. Various control devices that are usually provided in the power plants include fabric filters (FF) and electrostatic precipitators (ESP) for fly ash removal and flue gas desulfurization (FGD) and selective catalytic reactors (SCR) for SO_2 and NO_x removal, respectively. The efficiency of these devices for mercury removal depends on the speciation of mercury in the flue gas. Elemental mercury (Hg^0) is rather inert to most of these pollution control devices and requires to be converted to its oxidized form for its removal [25]. All the particulate control devices can easily remove mercury associated with fly ash and unburnt carbon while FGD and SCR as well as FF can remove or aid removal of oxidized mercury (Hg^{2+}) from flue gases. Coal blending and coal additives can be used to oxidize the flue gas mercury hence aiding mercury removal by these control devices. Special mercury control techniques such as activated carbon injection (ACI) uses activated carbon or activated carbon impregnated with chloride ions to aid oxidation of elemental mercury vapors on the AC surface. Effectiveness of any co-benefit or dedicated mercury removal technology will depend on the extent of oxidation of elemental mercury in the control devices. Latest regulations on pollution from thermal power plants have limited mercury emissions to 0.03 mg/Nm^3 for all facilities. However, limits for SO_2 and NO_x are limited to 100 mg/Nm^3 each for plants installed after 2017 and 60 mg/Nm^3 (< 500 MW capacity units)/200 mg/Nm^3 (>500 MW) for SO_2 along with 300 mg/Nm^3 for NO_x in case of plants established between 2003 and 2016, respectively [26]. Guidelines for mercury recently introduced in India is similar to the regulations in place in China; however, it is very high as compared to the USA where mercury limit for bituminous coal combustion facility is 0.0017 mg/Nm^3 and in case of lignite combustion is 0.0153 mg/Nm^3 [27]. Since Indian coal is characterized by low sulfur content and high ash content, particulate matter control devices commonly ESP is installed almost in all thermal power plants. Mercury removal options which co-benefit PM control is a good option for mercury emissions control from thermal power plants in India.

8.4.1 CO-benefit Removal Techniques

Electrostatic precipitators (ESP) which use electrostatic charges to separate dust particles from flue gas stream are useful for removing particulate mercury from flue gases. However, no oxidation of mercury occurs in ESP which results in passage of elemental mercury vapor from this device. A study conducted using coal from Talcher, Orissa, evaluated mercury removal performance of conventional and low-temperature ESP. The coal used had a mercury concentration of 220 g/kg and chlorine content of 100 mg/kg. Mercury removal rate in high-performance ESP with an inlet temperature of 90 °C showed a 68% mercury removal percent while conventional one only 19%. This temperature drop (up to 90 °C as compared to 170 °C in conventional ones) was brought about by gas–gas heat exchanger (GGH) [28]. Another study with SCR, coal side ESP, and wet FGD suggested a positive correlation between mercury oxidation in SCRs and removal in CS-ESPs [29]. This suggests that decreased temperatures in CS-ESP increase the efficiency of capture of particulate as well as oxidized mercury while it requires some other technologies to aid oxidation of mercury vapors. ESP alone showed a mercury removal efficiency of 28% for a study conducted in China [14]. Fabric filters which are another particulate control device consisting of cloth or fiber bags that retain fly ash and unburnt carbon on its pores are another device which showed a 67% mercury removal efficiency.

Selective catalytic reactor (SCR) is one of the few devices used for NO_x removal and the only one of them which facilitates mercury oxidation. It uses NH_3 as the reducing agent to reduce NO_x to N_2 using oxides of vanadium, molybdenum, and tungsten as the catalyst. This device in association with ESP and wet fluidized gas desulfurization (WFGD) can give a mercury removal efficiency of 69% and with the addition of fabric filters (FF), it increases to 90–95% [14]. A study done for mercury removal efficiency of SCR from flue gas from lignite burning suggests the important role played by the presence of halogen in coal. Oxidation in SCR occurs by two methods, homogenous oxidation and heterogeneous oxidation. Homogeneous oxidation occurs in the gaseous phase and heterogeneous oxidation occurs at the interface of solids. Since chloride content in lignite is less, the addition of various halogen gases and its effect on the oxidation of mercury in SCR is studied. For all the halogen gases (HCl, HF, HI, and HBr), mercury oxidation percentage increased with increasing the gas concentration up to a certain point after which it remained unchanged. Addition of HBr and HI showed stronger impact on Hg(0) oxidation than Hcl and HF. One more factor affecting the oxidation in SCR is the presence of NH_3. The reactions, NO_x reduction by NH_3 and Hg(0) oxidation by halogen species, compete with each other except when HBr and HI is used [30]. Gas flow rate, temperature of the flue gas, and presence of other gases are some of the parameters that affect Hg(0) oxidation in SCR. Positive correlation between Hg oxidation in SCRs and CS-ESPs as well as between SCR and wet FGDs indicates the importance of SCR in mercury removal from coal combustion flues gases [29]. Fluidized gas desulfurization (FGD) is used to scrub SO_2 from the flue gases. Two types of FGD are available, one dry FGD and other wet FGD. Water-soluble nature of oxidized Hg^{2+}

makes wet FGD a good device for oxidized mercury removal. However, it is necessary to have an increased Hg^{2+} concentration in flue gases prior to wet FGD. Oxidized mercury ions in wet FGD are prone to re-emission which was seen to be suppressed by addition of chemicals like Na_2S_4, NaHS, and HI. Also, increase in scrubber temperature and decreased pH of slurry due to increase in O_2 concentration further enhanced re-emission in scrubber [29]. Wet FGD with ESP present upstream shows a 64% mercury removal while FF installed upstream as particulate removal device increases the removal efficiency to 86%. Including a SCR upstream of particulate removal device further increases the removal efficiency to 69 and 90%, respectively.

8.4.2 Dedicated Removal Techniques

Activated carbon injection uses activated carbon as an adsorbent for mercury removal. Some of the parameters important while commercially tapping ACI are even distribution of sorbent in flue gases, sufficient contact time between sorbent and flue gases and maintaining system operating temperature within range. Standard PAC injection was effective for mercury capture in low sulfur bituminous coal but not in case of low rank fuel fitted with ESPs and those burning low rank fuel with FF as control equipment. This is so because such fuel shower lower chlorine levels. Also plants using high sulfur fuel showed no considerable effect of PAC injection [31]. ACI upstream of a particulate control device like ESP or FF alone gives a 90% removal efficiency [14]. TOXECON is a combination of ESP, ACI downstream of ESP, and pulse jet FF to capture the activated carbon. This system is considered to be more effective and more expensive than ESP/FF + ACI system [32]. This increased cost is to separate toxic fly ash and activated carbon from usual fly ash and unburnt carbon in primary PM control equipment usually ESP. TOXECON 2 addresses this problem without the additional cost of pulse jet FF by injecting PAC into the initial collecting field of ESP thus keeping the fly ash collected in the downstream collecting fields relatively free from toxic mercury [33]. Regeneration of used activated carbon using thermogravimetric analyzer and high-temperature air slide apparatus at a temperature around 900 F has been studied. A 46% Hg removal at 800 F and 100% at 900 F was observed. Mercury adsorption capacity of ACI increases with catalyst, 1% CuO and 4% Fe_2O_3 and 5% CuCl, and decrease in flue gas temperature. In case of flue gas with low halogen content, halogen-treated activated carbon is effective at capturing mercury. Presence of high sulfur levels in coal or increased SO_3 concentrations in flue gases hamper the removal efficiency of activated carbon. Separation of AC ash from fly ash becomes a major concern while using ACI technique, especially, when the ash is utilized in other boilers [32]. The study done on Talcher coal suggested that no effective mercury removal took place on activated carbon surface without the aid of denitrification catalysts [28]. In another study, ACI was used in conjunction with SCR, FF, and wet FGD and also with ESP in addition to achieve a removal efficiency of 97 and 99%, respectively. Activated carbon along with some advanced PM control technologies is believed to achieve removal efficiency as high as 90% [14]. Another study points out some of the limitations that come in the use of AC. The effectiveness of AC reduces in the presence of SO_2 and at high temperatures. Also, halogenation of AC is required prior to its use as an adsorbent. A high operating temperature has an adverse effect when plants with hot-side ESP are used. Few scenarios which limit the use of PAC for mercury removal are plants where complete removal of Hg was obtained at 700 F [34]. Choice of any pollution control measure depends not only on the removal efficiency but also on the cost incurred in the lifetime of the device. Presently, thermal power plants in India is equipped with only ESPs, a few of them do have FGD unit (Mumbai) [24]. Hence, a mercury removal system in Indian power plants should revolve around particulate matter control device. Among all the control devices that could act as co-benefit for mercury removal in power plants, wet FGD has highest capital cost and maintenance cost followed by SCR and FF which have around same capital cost though SCR has higher maintenance cost. ACI when introduced in a system consisting of SCR, FF and wet FGD would require an additional capital cost of around 10 CNY/kW and O&M cost of 11 CNY/kW/year [14]. While another study gives the capital cost of APC as $5/kW in the USA which is fairly low compared to other devices. Brominated carbon though higher in capital cost requires much lesser injection rate.

8.4.3 Coal Pre-treatment Techniques

Another option to control mercury in flue gases is coal washing and other coal pre-treatment, which reduces fly ash, sulfur, and mercury content of coal hence reducing the mercury emissions as well as increasing the efficiency of plant. Physical coal cleaning consists of reducing size of coal and screening, gravity separation of coal from sulfur-bearing mineral impurities, dewatering, and lastly drying. 10–50% Hg removal can be achieved by physical coal washing. The variation depends on the type of process used for cleaning and the nature of mercury in coal [33]. A conventional coal washing removes mercury associated with incombustible minerals and not the ones associated with organic carbon in the coal matrix [35]. Another study conducted to know mercury removal efficiency of coal washing and coal pre-treatment processes suggested 8–96% removal by coal cleaning and deshaling process while a 45–70% removal when treated at 300 °C. A combination of coal cleaning and thermal pre-treatment is ideal to achieve good mercury removal [36]. Coal beneficiation and coal blending are two additional treatments that can be provided to coal for enhanced mercury removal. *K*-Fuel which is widely used for sub-bituminous and lignite coal is a method of coal beneficiation which includes coal washing. It removes particulate matter, increases the heating value, and reduces emissions from the coal combustion [33]. Coal blending is commonly used to reduce SO_2 emissions from coal combustion. However, coal blending also provides an additional advantage of mercury removal. Coal blending increases chlorine in flue gases thus resulting in increased Hg removal. A mix of 60% sub-bituminous and 40% bituminous coal increased mercury oxidation to 63% in the absence of SCR and 97% with SCR. Some of the salts that are added to coal are hydrochloric acid (HCl) or ammonium chloride (NH_4Cl). These can be either sprayed on coal or injected in the boiler. It can also be added in the form of solid upstream of coal pulverizer [33, 35].

8.5 Summary

A total 1332.959 MT of total mercury has been emitted by coal combustion sector in India over the period of eight years starting 2005. The rate of emission is likely to increase further in the coming years with increase in coal consumption. Major coal consumption sector in India is thermal power plants consuming 71.56% of total coal available in India. Currently, the mercury emissions from thermal power plant are estimated between 19 and 130 g/Nm^3 [37]. Bituminous coal is the prevalent coal type available in thermal power plants in India. Thermal power plants being a major coal consumer in India require strict regulations, effective control devices, and continuous monitoring system to mitigate mercury emissions. Coal burned in India has high fly ash content and low sulfur and chloride content. Hence, approximately 60–70% of the mercury emitted in the flue gases is in elemental state and hence is emitted into the atmosphere. Almost all the power plants in India have ESP installed for PM control. Mercury removal efficiency of conventional ESP varies from 19 to 28%. However, the performance of cold-side ESP is seen to be better than the conventional one. ESP majorly removes particulate mercury from the flue gas which constitutes around 20–40% of the total mercury in the flue gases from thermal power plant. Removal of mercury in elemental vapor state and also oxidized mercury in flue gases require other treatment options which are currently not used in thermal power plants in India. Wet fluidized gas desulfurization (WFGD) removes oxidized mercury from flue gases as they are water soluble. However, WFGD has the highest capital and O&M cost among all the co-benefit control devices. Studies have shown a 64% mercury removal efficiency when a combination of ESP and WFGD is used. Fabric filters (FFs), another PM control device with a higher PM removal efficiency than ESP and also has a better mercury removal efficiency when compared to ESP. FF along with WFGD gives a mercury removal efficiency of 86%. Fabric filters though involving a higher capital investment has a lower O&M investment and quite high removal efficiency than ESP. Hence, replacement of ESP with FF can be considered with more stringent PM regulations introduced recently. Since only 30% of the mercury in flue gases exists in the form of oxidized mercury (Hg^{2+}) and rest 60–70% is in elemental form, it makes sense to supplement WFGD with control devices or techniques that would oxidize elemental mercury in flue gases hence improving the removal efficiency of WFGD. Selective catalytic reactor (SCR) is one such device placed upstream of PM control device, which reduces NO_x concentration in the flue gases also aiding the oxidation of elemental mercury. One study gives the removal efficiency of SCR, cold-side ESP, and WFGD combination as 69% which is just 5% increase from when ESP + WFGD was considered. SCR has an initial capital cost comparable with FF but a higher O&M cost. In India's context, use of SCR should not be considered unless

NO_x removal from thermal power plants is desired. Activated carbon injection (ACI) is a dedicated mercury removal technique which uses powered activated carbon or sometimes a halogenated activated carbon for removal of oxidized as well as elemental mercury from flue gases. Spent AC is then removed using a PM control device downstream. ACI provided upstream of an ESP or FF showed a mercury removal efficiency of 90%. ACI requires less capital and O&P cost according to the study conducted in USA and China; however, costs involved in India still requires to be studied. TOXECON is an advancement in ACI mercury removal, which provides an ESP upstream and a pulse jet FF downstream of ACI. This separates usual fly ash from toxic AC and high-mercury fly ash. Taking into consideration the efficiency and costs involved, a TOXECON or halogenated ACI + PM control seems to be a good option for mercury removal in Indian thermal power plants. Coal cleaning and blending with additives are also cheap and effective techniques to reduce mercury as well as sulfur impurities from coal. Study of efficiency of coal washing for mercury removal in case of Indian coal varied from around 15 to 30%. Implementation of coal washing should be considered based on the nature of coal mined from different regions. Also, costs involved in handling of effluents released from washing should also be considered when using this technique. One of the most important steps to ensure mercury emissions reduction from coal combustion is to put a transparent emission monitoring system in place.

References

1. Risher, J.F., World Health Organization.: Elemental mercury and inorganic mercury compounds: human health aspects
2. Schroeder, W.H., Munthe. J.: Atmospheric mercury-an overview. Atmos. Environ. **32**(5) 809–822 (1998). Elsevier, Great Britain
3. Pirrone, N.: Emissions, Hemispheric Transport of Air Pollution 2010, pp 75–96. United Nations Economic Commission for Europe
4. UNEP., WHO.: Guidance for identifying population at risk from mercury exposure (2008)
5. UNEP.: Minamata Convention on Mercury (2013)
6. Chen.Y., Wang, R., Shen, H., Li, W., Chen, H., Huang, Y., Zhang, Y.: Global mercury emissions from combustion in light of international fuel trading. Environ. Sci. Technol. **48**, 1727–1735 (2014)
7. Berg, T., Bartnicki, J., Munthe, J., Lattila, H., Hrehoruk, J., Mazur, A.: Atmospheric mercury species in the Europian Arctic: measurements and modelling. Atmos. Environ. **35**, 2569–2582 (2001). Elsevier
8. Chen, L., Wang, H.H., Liu, J.F., Tong, Y.D., Ou, L.B., Zhang, W., Hu, D., Chen, C., Wang, X.J.: Intercontinental transport and deposition patterns of atmospheric mercury from anthropogenic emissions. Atmos. Chem. Phys. **14**, 10163–10176 (2014)
9. Travnikov, O.: Contribution of the intercontinental atmospheric transport to mercury pollution in the Northern Hemisphere. Atmos. Environ. **39**, 7541–7548 (2004). Elsevier
10. United Nations Environment Programme.: Toolkit for Identification and Quantification of Mercury Releases Version 1.4 (2017)
11. Pacyna, E.G., Pacyna, J.M.: Global Emission of Mercury from Anthropogenic Sources in 1995 (2001)

12. Pacyna, E.G., Pacyna, J.M., Sundseth, K., Munthe, J., Kindbom, K., Wilson, S., Steenhuisen, F., Maxson, P.: Global emission of mercury to the atmosphere from anthropogenic sources in 2005 and projections to 2020. Atmos. Environ. **44**, 2487–2499 (2010)
13. Mukherjee, A.B., Bhattacharya, P., Sarkar, A., Zevenhoven, R.: Mercury emissions from industrial sources in India and its effects in the environment. In: Mason, R., Pirrone, N. (eds.) Mercury Fate and Transport in the Global Atmosphere. Springer, Boston, MA (2009)
14. Ancora, M.P., et al.: Economic analysis of atmospheric mercury emission control for coal-fired power plants in China, pp 1-11. JES, Elsevier B.V (2015)
15. International Energy Agency.: https://www.iea.org/
16. Coal Controller's Organisation, GOI, Ministry of Coal, Provisional Coal Statistics, 2005–06
17. Coal Controller's Organisation, GOI, Ministry of Coal, Provisional Coal Statistics 2006–07
18. Coal Controller's Organisation, GOI, Ministry of Coal, Provisional Coal Statistics 2007–08
19. Coal Controller's Organisation, GOI, Ministry of Coal, Provisional Coal Statistics 2008–09
20. Coal Controller's Organisation, GOI, Ministry of Coal, Provisional Coal Statistics 2009–10
21. Coal Controller's Organisation, GOI, Ministry of Coal, Provisional Coal Statistics 2010–11
22. Coal Controller's Organisation, GOI, Ministry of Coal, Provisional Coal Statistics 2011–12
23. Coal Controller's Organisation, GOI, Ministry of Coal, Provisional Coal Statistics 2012–13
24. Prepared, R.: Assessment of the mercury content in coal fed to power plants and study of mercury emissions from the sector in India (2014)
25. UNEP.: Guidance on best available techniques and best environmental practices to control (2015)
26. The Gazette of India.: Ministry of Environment, Forest and Climate Change Notification. Mercury Emissions from Coal-fired Power Plants and Coal-fired Industrial Boilers', pp. 1–45 (2016)
27. NRDC.: Health facts summary of recent mercury emission limits for power plants in the united (2012)
28. Japan Coal Energy Centre.: Environmental Study through combustion test of Indian coal (2017)
29. Pudasainee, D., et al.: Oxidation, reemission and mass distribution of mercury in bituminous coal-fired power plants with SCR, CS-ESP and wet FGD. Fuel **93**(2012), 312–318 (2012). Elsevier Ltd
30. Cao, Y., et al.: Impacts of halogen additions on mercury oxidation, in a slipstream Selective Catalyst Reduction (SCR), reactor when burning sub-bituminous coal. Environ. Sci. Technol. **42**(1), 256–261 (2008)
31. Srivastava, R.K., et al.: Control of mercury emissions from coal-fired electric utility boilers. Environ. Sci. Technol. **40**(5), 1385–1393 (2006). https://doi.org/10.1021/es062639u
32. Sjostrom, S., et al.: Activated carbon injection for mercury control: overview. Fuel **89**(6), 1320–1322 (2010). Elsevier Ltd
33. Pacyna, J.M., et al.: An assessment of costs and benefits associated with mercury emission reductions from major anthropogenic sources. J. Air Waste Manag. Assoc. **60**(3), 302–315 (2010)
34. Okwadha, G.D.O., et al.: Thermal removal of mercury in spent powdered activated carbon from TOXECON Process, pp 1032–1040 (2009)
35. UNEP Chemicals Branch., Geneva., S.: Process optimization guidance for reducing mercury emissions from coal combustion in power plants, p. 82 (2010)
36. Dziok, T., Strugała, A.: Method selection for mercury removal from hard coal 2007 (2017)
37. Centre for Science and Environment.: Clearing the air: pollution-control technology for coal-based power plants (2016)

Chapter 9
Saponin-Aided Reverse Micellar Extraction of Malachite Green Dye From Aqueous Solutions

R. Subashini, N. Sivarajasekar, K. Balasubramani and J. Prakashmaran

Abstract Synthetic dyes used for the dyeing process increase the rate of water body pollution due to the discharge from the industries. There are various physical, chemical, and biological methods available to treat dye bearing effluents. Reverse micellar extraction (RME) is one of the physical method and will be an efficient method if there are cost-effective amendments. Surfactant-assisted solvent extraction is proven technology for protein and enzyme extraction. In this study, saponin is extracted from *Sapindus mukorossi* nuts. Saponin is characterized by FTIR. Malachite green is taken as model pollutant. Extraction of dye from aqueous solution is carried out at various conditions such as saponin concentration (1–3 mg/L), dye concentration (100–200 mg/L), and solvent ratio (1:1–1:3). The process parameters are optimized via Box–Behnken RSM design. The optimal values are predicted using Derringer's desirability.

Keywords Extraction · Malachite green · Saponin · RSM · Reverse micellar extraction

9.1 Introduction

Industrial wastewater contains many organic and inorganic substances which are nondegradable. The textile industry effluent which is having different dyes is the major water pollutant throughout India [1–3]. Their nondegradability affects the growth of phytoplanktons, zooplanktons, and living organisms, produces odor, and changes

---

R. Subashini · N. Sivarajasekar (✉)
Laboratory for Bioremediation Research, Unit Operations Laboratory, Department of Biotechnology, Kumaraguru College of Technology, Coimbatore 641049, India
e-mail: sivarajasekar@gmail.com

K. Balasubramani
Department of Petrochemical Engineering, JCT College of Engineering and Technology, Coimbatore 641105, India

J. Prakashmaran
Department of Food Science and Nutrition, Periyar University, Salem 636011, India

© Springer Nature Singapore Pte Ltd. 2020

V. Sivasubramanian and S. Subramanian (eds.), *Global Challenges in Energy and Environment*, Lecture Notes on Multidisciplinary Industrial Engineering, https://doi.org/10.1007/978-981-13-9213-9_9

the color of the water body. For humans, it causes respiratory disease, skin allergies, and hemorrhages and makes the place uncomfortable to live [4, 5]. Generally, the reduction of the toxic compounds from the effluents before discharge was done by the effluent treatment which is classified into primary, secondary, and tertiary treatments [6–8]. Traditional methods such as adsorption, catalytic degradation, precipitation, coagulation, bioremediation, and membrane separation are used to remove the chemicals of both organics and inorganics; however, each method has its demerits [9–15]. Solvent extraction is one of the kinds to extract the pollutant from the wastewater which includes reverse micellar extraction (RME) [16]. Reverse micelles are thermodynamically stable; nanometer-sized water droplets dispersed in an organic phase by means of surfactant. The RME is a proven efficient method for bioproducts recovery [17]. The RME uses the surfactant which acts as a primary carrier to pull out the pollutant to the solvent phase. The surfactant is used to increase the emulsification and it improves the efficiency of mass transfer. However, the commercially available surfactants are costly and harmful due to recalcitrant nature. The alternative way is utilizing the naturally available biosurfactants. Saponin is one of the biosurfactant which is derived from the plant source, microorganisms, animals, and humans. The saponin derived from the plant source is nontoxic and biodegradable which can replace the use of chemical surfactant. In addition, the presence of saponin does not affect the quality of the water. The common source for the saponin is the soapnut (*Sapindus mukorossi*) which is abundantly present in India [18]. The saponin can be obtained from various seeds like chickpeas, soya beans, sprouts, soapnut, and etc. [19, 20]. The objective of this article is to extract saponin from soapnut and utilize it for RME. Further to optimize the RME parameters such as temperature, solvent ratio, and saponin loading to increase the RME efficiency.

9.2 Materials and Methods

9.2.1 Extraction of Saponin from Soapnut Powder and its Characterization

About 10 g of soapnut powder was added with 100 ml of ethyl acetate in a 250 ml beaker. The beaker was placed in a water bath at 60 °C with intermittent stirring for 7 h. After heating, it allows to cool till it attains the room temperature. Then the solution was filtered using the Whatman filter paper no. 1 and the solvent was recovered by the rotary flash evaporator. The sample was concentrated and placed in a Petri plate under 27 °C for drying. Finally, the sample was weighed for calculating the yield of surfactant. The extracted saponin was characterized using FTIR (IRAFFINITY-1S, Double beam, Shimadzu, US) in the range of 750–4500 cm^{-1}. The critical micellar concentration of saponin was determined using a procedure explained by [21].

9.2.2 Optimization of RME Parameters on Dye Removal

Addition of surfactant to the aqueous solvent mixture induces the reverse micellar formation. Constant volume of malachite green dye aqueous solution and the isoamyl alcohol were mixed in different ratios (1:1–1:3) to analyze the solvent ratio effects. The initial concentration of the aqueous malachite green solution was altered between 100 and 200 mg/L and experiments carried out to understand the concentration effects. Saponin loading also varied between 1 and 3 mg/L mixture to investigate the surfactant loading effects. The entire optimization is carried out using Box–Behnken design. Design-Expert 10 (State Ease, USA) software is used to interpret the results. In each case, aqueous phase was taken and analyzed using UV spectrophotometer at 620 nm. Experiments were conducted in random order to reduce error. The regression analysis was executed to evaluate the response function. The following quadratic model Eq. (9.1) can predict the extraction efficiency (%E):

$$\%E = \beta_0 + \sum_{i=1}^{k} \beta_i X_i + \sum_{j=1}^{k} \beta_i X_i^2 + \sum_{i<j}^{k} \beta_{ij} X_{ij} + \cdots \quad (9.1)$$

where %E is the predicted response (Extraction efficiency), β_i, β_j, and β_{ij} are the coefficients from regression analysis, k is the number of factors, X_i and X_j represent the uncoded value of the ith (linear coefficient) and jth (quadratic coefficient) parameter, respectively [22].

9.3 Results and Discussion

9.3.1 Characterization of Saponin

The saponin yield was obtained as 26 g/kg powder of dried soapnut powder. The critical micellar concentration of the extracted saponin was found to be 0.2 g/L ($=4.37 \times 10^{-4}$ mol/L). The FTIR results clearly indicated that the extract is saponin which is known from the set reference. The XRD results also indicated that the extracted saponin powder is crystalline and polar in nature.

9.3.2 Statistical Analysis of RME

Response surface methodology (RSM) is more advantageous than the traditional single parameter optimization because it can save time, space, and raw material. In experimental design, a BBD is a type of RSM. BBD along with the design of experiments removes systematic errors with an estimate of the experimental errors and also reduces the number of experiments in order to obtain the optimum operating conditions [3, 23]. Three factors, namely solvent ratio (1:1–1:3 mg/L, X_1), initial dye concentration (100–200 mg/L, X_2), and concentration of surfactant (1–3 mg/L, X_3) were considered to analyze the extraction efficiency of malachite green dye from aqueous solution. The results for this design are shown in Table 9.1. Multivariate regression analysis of the empirical data generated the following quadratic equation:

$$\begin{aligned}\%E =&70.95 - 4.76 \times X_1 - 0.32 \times X_2 + 14.53 \times X_3 - 0.82 \times X_1 \times X_2 \\ &+ 2.17 \times X_1 \times X_3 - 0.04 \times X_2 \times X_3 + 3.61 \times X_1^2 + 2.05 \times X_2^2 - 4.21 \times X_3^2\end{aligned} \quad (9.2)$$

The extraction efficiency values predicted from the Eq. (9.2) were almost close to the linearity in the normal probability plot (Fig. 9.1) indicated the close fit. Hence, this quadratic polynomial equation is well fit for this batch experiments. Nevertheless, the association and the suitability of this model equation were established by using analysis of variance (ANOVA) test. ANOVA results are presented in Table 9.2.

Probability less than 0.05 signifies the model coefficients are important. From Table 9.2, factors (X_1, X_2, X_3) and their interactions (X_1X_2, X_1X_3, X_1^2, X_2^2, X_3^2) were noteworthy. The small value of probability of the model F value (16985.46) indicated that the selected quadratic polynomial model is significant. The value of regression coefficient (0.99), close to unity, showed that predicted model coefficients are reasonable. The predicted R^2 (0.998) and adjusted R^2 (0.996) are in practical

Table 9.1 BBD matrix for extraction efficiency of malachite green

X_1	X_2	X_3	$\%E_{exp}$	$\%E_{pred}$
1	−1	0	71.35	71.36
−1	1	0	82.46	82.51
0	0	0	72.41	72.36
0	0	0	80.26	80.25
1	1	0	53.12	53.24
1	0	−1	58.33	58.41
0	−1	−1	78.04	77.96
−1	0	1	91.94	91.82
−1	0	−1	54.67	54.54
0	0	0	54.06	53.99
0	1	−1	83.62	83.69
1	0	1	82.83	82.96
−1	−1	0	70.96	70.95
0	0	0	70.94	70.95
0	0	0	70.97	70.95
0	−1	1	70.95	70.95
0	1	1	70.95	70.95

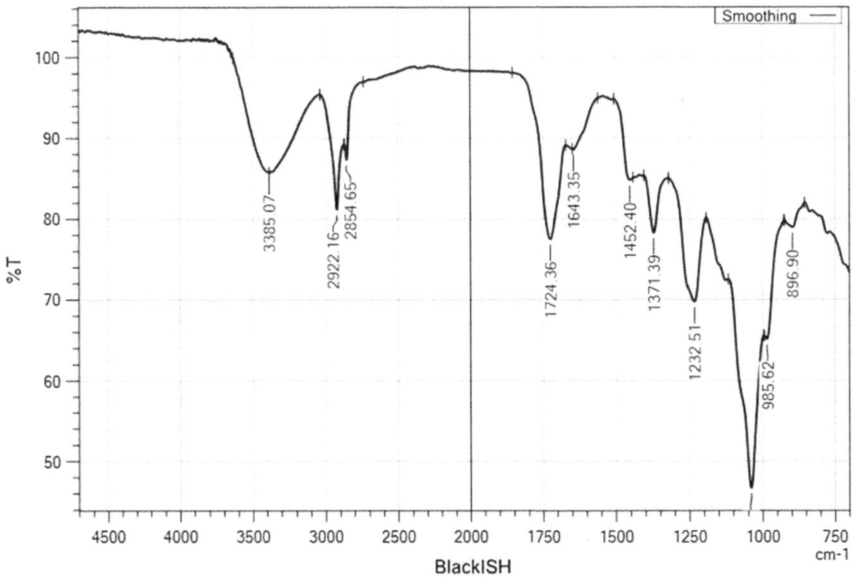

Fig. 9.1 FTIR analysis of saponin

Table 9.2 ANOVA results

Source	Sum of squares	df	Mean square	F value	p-value Prob > F
Model	2033.46	9	225.94	16985.5	< 0.0001
X_1	181.166	1	181.166	13619.5	< 0.0001
X_2	0.80963	1	0.80963	60.8653	0.0001
X_3	1689.4	1	1689.4	127004	< 0.0001
$X_1 X_2$	2.6569	1	2.6569	199.737	< 0.0001
$X_1 X_3$	18.879	1	18.879	1419.27	< 0.0001
$X_2 X_3$	0.00766	1	0.00766	0.57557	0.4728
X_1^2	55.0203	1	55.0203	4136.25	< 0.0001
X_2^2	17.7142	1	17.7142	1331.7	< 0.0001
X_3^2	74.6765	1	74.6765	5613.95	< 0.0001
Residual	0.09311	7	0.0133		
Lack of fit	0.09259	3	0.03086	237.42	0.5100
Pure error	0.00052	4	0.00013		
Cor total	2033.55	16			

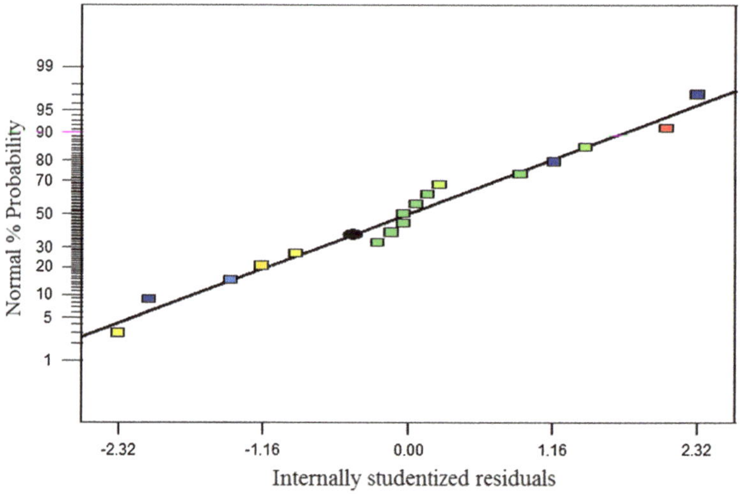

Fig. 9.2 Predicted extraction efficiency versus actual values

tuning with each other. Greater than 4 of the adequate precision values specify an adequate signal. Hence, the predicted quadratic models could navigate the design space to understand the factor interactions (Fig. 9.2).

9.3.3 Significant Interactions of Parameters

Contour and 3D plots were produced to get a clear picture about the interactions of the independent parameters and optimum conditions for the dyes extraction. The nonlinear nature of the contours suggested that the interactions between the parameters are significant and the optimum parameters cannot be found simply. Each 3D plot characterizes the combinations of the two selected parameters and the shaded surface indicates the maximum extraction efficiency.

Effect of solvent ratio: From Fig. 9.3 a and b, as the solvent volume increases to the constant dye concentration, there is also an increase in the micellar formation which readily transferred the dye to the solvent phase [17]. So the concentration of

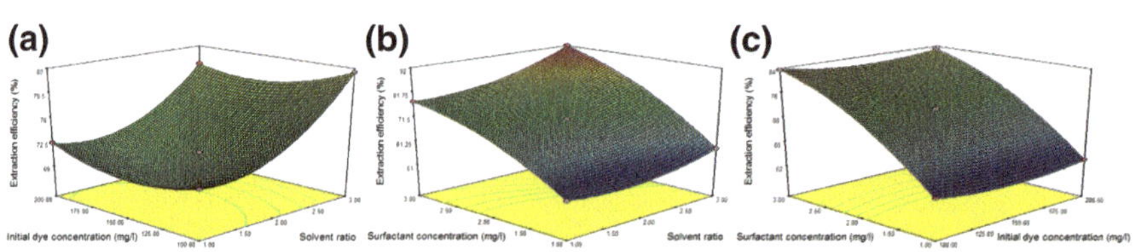

Fig. 9.3 Effect of process parameters of malachite extraction

dye compounds present in the solvent phase will be more than the aqueous phase. The optical density of the dye solution was measured at 620 nm and 88.92% of dye removed from the aqueous solution.

Effect of biosurfactant concentration: The saponin acted as an extractant and emulsifier. Saponin has the amphiphilic nature, so it attracts all the contaminants toward itself to form the micellar in which it enhances the transfer rate. From Fig. 9.3b and c, as the concentration of saponin increases, this readily improves the emulsion stability. When there is an increase in the surfactant concentration, the rate of extraction increases.

Effect of initial dye concentration: It is inferred from Fig. 9.3a and c that, when the initial dye concentration increases, it affects the mass transfer rate positively. The higher initial concentration enhanced the rate of transfer, subsequently resulted in increased extraction efficiency.

9.3.4 Optimization Using Derringer's Desirability Function

RME process parameters were optimized using Derringer's desirability function [24–28]. This function optimizes the response at selected parameter level by satiating the needs in the BBD design. A minimum surfactant concentration, minimum solvent ratio, and maximum dye concentration were fixed to optimize maximum extraction efficiency. By seeking from 12 starting points in the optimum results, the best local maximum for malachite dye extraction was found to be at solvent ratio (1:1), surfactant concentration (1.8 mg/L), and initial malachite concentration (100 mg/L) and extraction efficiency of 72.75% at 89% desirability. For validating the optimal values, confirmatory experiments were done at the optimized parameters and the extraction efficiency was found to be 72.85%. The obtained values were associated to the calculated data, indicating BBD joined with desirability function could be efficiently employed to optimize the batch extraction process parameters of the malachite green dye.

9.4 Conclusions

Malachite green dye from aqueous phase was extracted using reverse micellar extraction technique. Saponin extracted from soapnut was utilized as emulsifier and carrier. Batch experiments were carried out and the extraction efficiency was investigated by using amyl alcohol as solvent. The process parameters such as dye concentration, sulphuric acid concentration, and saponin concentration were optimized using BBD. The effects of parameters were studied by using 3D response surface plots. The optimum conditions were calculated by using Derringer's desirability function and were found to be at solvent ratio (1:1), surfactant concentration (1.8 mg/L), and initial malachite concentration (100 mg/L).

References

1. Karthik, V., Saravanan, K., Sivarajasekar, N., Suriyanarayanan, N.: Bioremediation of dye bearing effluents using microbial biomass. Ecol. Environ. Conserv **22**, S423–S434 (2016)
2. Karthik, V., Saravanan, K., Sivarajasekar, N., Suriyanarayanan, N.: Utilization of biomass from Trichoderma harzianum for the adsorption of reactive red dye. Ecol. Environ. Conservation **22**, S435–S440 (2016)
3. Sivarajasekar, N., Balasubramani, K., Mohanraj, N., Maran, J.P., Sivamani, S., Koya, P.A., Karthik, V.: Fixed-bed adsorption of atrazine onto microwave irradiated *Aegle marmelos* Correa fruit shell: Statistical optimization, process design and breakthrough modeling. J. Mol. Liq. **241**, 823–830 (2017)
4. Karthik, V., Sivarajasekar, N., Padmanaban, V.C., et al.: Biosorption of xenobiotic reactive Black B onto metabolically inactive *T. harzianum* biomass: optimization and equilibrium studies. Int. J. Environ. Sci. Technol. (Article under press) (2018). https://doi.org/10.1007/s13762-018-1841-5
5. Sivarajasekar, N., Baskar, R., Ragu, T., Sarika, K., Preethi, N., Radhika, T.: Biosorption studies on waste cotton seed for cationic dyes sequestration: equilibrium and thermodynamics. Appl. Water Sci. **7**(4), 1987–1995 (2017)
6. Selvaraj, K., Sevugaperumal, R., Ramasubramanian, V.: Impact of match industry effluent on growth and biochemical characteristics of *Cyamopsis tetragonoloba* Taub and amelioration of the stress by seaweed treatment. Indian J. Fund. Appl. Life Sci. 1–7 (2013)
7. Sivarajasekar, N.: Biosorption of Cationic Dyes Using Waste Cotton Seeds (Doctoral dissertation, Ph. D. Thesis. Anna University Chennai) (2014)
8. Sivarajasekar, N., Mohanraj, N., Balasubramani, K., Maran, J.P., Moorthy, I.G., Karthik, V., Karthikeyan, K.: Optimization, equilibrium and kinetic studies on ibuprofen removal onto microwave assisted–activated Aegle marmelos correa fruit shell. Desalination Water Treatment **84**, 48–58 (2017)
9. Ghaly, A.E., Ananthashankar, R., Alhattab, M.V.V.R., Ramakrishnan, V.V.: Production, characterization and treatment of textile effluents: a critical review. J Chem. Eng. Process Technol. **5**(1), 1–18 (2014)
10. Muthusaravanan, S., Sivarajasekar, N., Vivek, J.S., et al.: Phytoremediation of heavy metals: mechanisms, methods and enhancements. Environ. Chem. Lett. (Article under press) (2018). https://doi.org/10.1007/s10311-018-0762-3
11. Sivarajasekar, N.: Hevea brasiliensis-A Biosorbent for the adsorption of Cu (II) from aqueous solutions. Carbon Lett. **8**(3), 199–206 (2007)
12. Sivarajasekar, N., Mohanraj, N., Sivamani, S., Moorthy, G.I.: Response surface methodology approach for optimization of lead (II) adsorptive removal by *Spirogyra sp.* biomass. J. Environ. Biotechnol. Res. **6**(1), 88–95 (2017d)
13. Sivarajasekar, N., Mohanraj, N., Sivamani, S., Moorthy, I. G., Kothandan, R., Muthusaravanan, S.: Comparative modeling of fluoride biosorption onto waste *Gossypium hirsutum* seed microwave-bichar using response surface methodology and artificial neural networks. In: Intelligent Computing, Instrumentation and Control Technologies (ICICICT), 2017 International Conference on IEEE, pp. 1631–1635 (2017e)
14. Sivarajasekar, N., Paramasivan, T., Muthusaravanan, S., Muthukumaran, P., Sivamani, S.: Defluoridation of water using adsorbents-a concise review. J. Environ. Biotechnol. Res. **6**(1), 186–198 (2017)
15. Sivarajasekar, N., Paramasivan, T., Subashini, R., Kandasamy, S., Prakash Maran, J.: Central composite design optimization of fluoride removal by spirogyra biomass. Asian J. Microbiol. Biotechnol. Environ. Sci. **19**, S130–S137 (2017)
16. Sivarajasekar, N., Ramasubbu, S., Maran, J. P., Priya, B.: Cationic dyes sequestration from aqueous phase using biosurfactant based reverse micelles. In: Recent Advances in Chemical Engineering pp. 67–74. Springer, Singapore (2016)
17. Pandit, P., Basu, S.: Removal of ionic dyes from water by solvent extraction using reverse micelles. Environ. Sci. Technol. **38**(8), 2435–2442 (2004)

18. Sarin, J.L., Beri, M.L.: Extraction of saponin from soap nut. Ind. Eng. Chem. **31**(6), 712–713 (1939)
19. Edeoga, H.O., Okwu, D.E., Mbaebie, B.O.: Phytochemical constituents of some Nigerian medicinal plants. Afr. J. Biotech. **4**(7), 685–688 (2005)
20. Fenwick, D.E., Oakenfull, D.: Saponin content of food plants and some prepared foods. J. Sci. Food Agric. **34**(2), 186–191 (1983)
21. Chakraborty, T., Chakraborty, I., Ghosh, S.: The methods of determination of critical micellar concentrations of the amphiphilic systems in aqueous medium. Arab. J. Chem. **4**(3), 265–270 (2011)
22. Sivarajasekar, N., Mohanraj, N., Baskar, R., Sivamani, S.: Fixed-bed adsorption of ranitidine hydrochloride onto microwave assisted—Activated *aegle marmelos correa* fruit shell: Statistical Optimization and Breakthrough Modelling. Arab. J. Sci. Eng. **43**(5), 2205–2215 (2018)
23. Sivarajasekar, N., Mohanraj, N., Sivamani, S., Prakash Maran, J., Ganesh Moorthy, I., Balasubramani, K.: Statistical optimization studies on adsorption of ibuprofen onto Albizialebbeck seed pods activated carbon prepared using microwave irradiation. Mater Today Proc. **5**, 7264–7274 (2018)
24. Janani, K., Sivarajasekar, N., Muthusaravanan, S., Ram, K., Prakashman, J., Sivamani, S., Dhakal, N., Shahnaz, T. Selvaraju, N.: Optimization of EDTA enriched phytoaccumulation of zinc by Ophiopogon japonicus: comparison of response surface, artificial neural network and random forest models. Bioresour. Technol. Rep. **7**, 100265. https://doi.org/10.1016/j.biteb.2019.100265
25. Muthusaravanan, S., Priyadharshini, S.V., Sivarajasekar, N., Subashini, R., Sivamani, S., Dharaskar, S., Dhakald, N.: Optimization and extraction of pharmaceutical micro-pollutant-norfloxacin using green emulsion liquid membranes. Desalin. Water Treat. **156**, 238–244 (2019). https://doi.org/10.5004/dwt.2019.23833
26. Sivarajasekar, N., Balasubramani, K., Baskar, R., Sivamani, S., Moorthy, I.G.: Eco-friendly acetaminophen sequestration using waste cotton seeds: equilibrium, optimization and validation studies. J. Water Chem. Technol. **40**(6), 334–342 (2018)
27. Sivarajasekar, N., Subashini, R., Devi, R. S.: Optimization of extraction methods for natural pigment from Lawsonia inermis. Int. J. Green Pharm. (IJGP). **12**(03), 728–732 (2018)
28. Sivarajasekar, N., Baskar, R.: Optimization, equilibrium and kinetic studies of basic red 2 removal onto waste gossypium hirsutum seeds. Iran. J. Chem. Chem. Eng. **37**(2), 155–167 (2018)

Chapter 10
Modeling, Simulation, and Parameter Estimation of Antisolvent Crystallization of α-Lactose Monohydrate

Nitin Pawar, Ravi Methekar and Shailesh Agrawal

Abstract The current work investigates the study of semi-batch antisolvent crystallization of α-lactose monohydrate. The antisolvent crystallization has a number of advantages over conventional evaporative crystallization like lower energy consumption and better control over morphology and a crystalline phase. Experiments were performed for the different initial concentration of lactose, i.e., at 0.25, 0.30, and 0.37 kg/kg of solution. Average crystal size and residual lactose concentration were analyzed using image analysis and refractive index, respectively. The mathematical model of the semi-batch antisolvent crystallization process was developed and simulated using MATLAB to predict the experimental results. A comparison of the experimental with the simulated results has been reported. The crystallization kinetic parameters, i.e., nucleation and growth parameters (k_g, g, k_b, and b) were estimated using $f_{\text{minsearch}}$ subroutine by minimizing the errors between model predicted and the experimental values.

Keywords Lactose · Antisolvent · Morphology · Modeling · PBM

Nomenclature

B	Nucleation rate (number/m^{3}s)
G	Growth rate (m/s)
b	Nucleation order
g	Growth order
k_b	Nucleation coefficient (number/m^{3}s)
k_g	Growth coefficient (m/s)
n	Population density of crystals (number/m^{3}s)
C	Lactose (solute) concentration (kg/kg of solution)

N. Pawar · R. Methekar · S. Agrawal (✉)
Department of Chemical Engineering, Visvesvaraya National Institute of Technology, Nagpur, India
e-mail: sgagrawal28783@gmail.com

© Springer Nature Singapore Pte Ltd. 2020

V. Sivasubramanian and S. Subramanian (eds.), *Global Challenges in Energy and Environment*, Lecture Notes on Multidisciplinary Industrial Engineering, https://doi.org/10.1007/978-981-13-9213-9_10

C_{eq}	Saturation concentration of lactose (kg/kg of solution)
L	Length of crystal (m)
k_v	Shape factor
S	Slurry density (kg/kg)
M	Mass (kg)
Q	Antisolvent addition rate (m^3/s)
x	Fraction of antisolvent
t	Time (s)
ρ	Density (kg/m^3)
i	Bin size

Subscripts

as	Antisolvent
cr	Crystal
s	Solution
i	Initial

10.1 Introduction

Lactose is the major dissolved component of milk whey and a byproduct of the milk industry [1]. It is also an important commodity compound in the food and pharmaceutical industry. It is widely used as an excipient and sweetener [2], and due to this, it has been of great academic and industrial interest. It generally occurs in α- and β-forms and exhibits tomahawk and needle/rod-shaped morphology [3]. The most stable form of lactose is the hydrated form, α-Lactose monohydrate. α-Lactose, when dissolved in aqueous solution, undergoes mutarotation and yields β-lactose. At equilibrium, β-lactose is present approx 60% wt/wt in solution [4].

Industrial lactose production involves concentration and evaporative crystallization of solution [5] and it is an energy-intensive process. As the worldwide demand for energy is growing, antisolvent crystallization has been suggested as an alternative to save stringent energy requirements, as this process can be carried out at the room temperature. Antisolvent crystallization can also be used for controlling nucleation, growth, polymorphism, and ultimately desired crystal size distribution. The added antisolvent can also be recovered and reused but no author has reported recovery and reuse of antisolvent in the crystallization process. Literature reported crystallization of lactose includes continuous crystallization [6, 7], antisolvent crystallization using acetone [8, 9], ethanol [10, 1], isopropanol [2], and sonocrystallization [11–14].

Nowadays, to describe the performance of the process and to obtain desired crystal size distribution, development of the effective mathematical model is an important issue. Nowee et al. [16] reported antisolvent crystallization model, simulation, and experimental validation, and later, Gogate et al. developed the same model to compare the simulated results with the experimental results [15] for benzoic acid.

All the previous work on antisolvent crystallization of lactose was conducted in batch mode. This work investigates semi-batch antisolvent crystallization of lactose. The effect of initial lactose concentration on crystal size and morphology has been studied. A mathematical model (discussed below) was also developed for antisolvent crystallization of lactose. The kinetic parameters were estimated by fitting the measured values to the values predicted by the model.

10.2 Mathematical Modeling

The constitutive modeling equations for antisolvent crystallization were taken from previous studies. The equations that needed to be solved simultaneously include population balance, solute balance, and the growth and nucleation kinetics. The assumptions in the population balance equation shown below are (a) size-independent growth, and (b) negligible agglomeration and breakage [15].

$$\frac{\partial n(L)}{\partial t} - B + \frac{G\partial n(L)}{\partial L} + \frac{n(L)\mathrm{d}M_s}{M_s\mathrm{d}t} = 0 \tag{10.1}$$

The general mass balance of this antisolvent crystallization process was adopted from Nowee et al. [16] and is given below:

$$\frac{\mathrm{d}M_s}{\mathrm{d}t} = \frac{\mathrm{d}M_c}{\mathrm{d}t} + \frac{\mathrm{d}M_L}{\mathrm{d}t} \tag{10.2}$$

The change in liquid mass is the addition of the initially added solvent and the antisolvent mass added during the whole process.

$$M_L = M_{i+}Q\rho_{\mathrm{as}}t \tag{10.3}$$

The change in solute (lactose) mass (M_c) has been represented as the mass of the crystals formed in the crystallizer due to nucleation and/or growth

$$M_c = \rho_{\mathrm{cr}}k_v \int n(L)L^3\mathrm{d}L \tag{10.4}$$

This change in solute mass Eq. (10.5) indicates that the rate at which dissolved lactose concentration decrease in the solution is equal to the rate of generation of the solid crystals in the solution.

$$\frac{\mathrm{d}C}{\mathrm{d}t} + \frac{\mathrm{d}S}{\mathrm{d}t} = 0 \tag{10.5}$$

The rate of generation of solute crystals can be represented using the following equation

$$\frac{\mathrm{d}C}{\mathrm{d}t} = -3\rho_c G k_v \int_{\infty}^{0} n(L)L^2 \mathrm{d}L \tag{10.6}$$

In the above equation, growth rate G (m/s) is given by the following equation

$$G = k_g(C - C_{\mathrm{eq}})^g \tag{10.7}$$

where $(C–C_{\mathrm{eq}})$ is the concentration driving force for crystallization to occur.

The primary nucleation rate (B) of the crystals is given as

$$B = k_b(C - C_{\mathrm{eq}})^b \tag{10.8}$$

Saturation concentration (C_{eq}) of the α-lactose monohydrate in ethanol–water mixture was determined experimentally at 40 °C and equation can be obtained by data fitting.

$$C_{\mathrm{eq}} = -0.70x^4 + 127.7x^3 - 34.34x^2 - 49.42x + 25.196 \tag{10.9}$$

where x can be represented as

$$x = \frac{Q\rho_{\mathrm{as}}t}{M_i + Q\rho_{\mathrm{as}}t} \tag{10.10}$$

Method of moments (MOMs) was used to solve the PBE by converting the PBE into series of ODEs that can be solved together with the mass balance equations [17].

The kinetics of this semi-batch antisolvent crystallization process, i.e., nucleation and growth parameters (k_g, g, k_b, and b) were estimated by using the $f_{\mathrm{minsearch}}$ subroutine in MATLAB, 2009. The objective function was to minimize the errors between model predicted and the experimental values for average crystal size and residual lactose concentration.

10.3 Materials and Methods

α-Lactose monohydrate (99% purity) was purchased from Scope Ingredients Pvt Ltd, Chennai, India. Ethanol was purchased from Merck India Pvt Ltd. The schematic of the experimental setup is shown in Fig. 10.1. It consists of a jacketed crystallizer of 500 ml capacity, thermostatic water circulator for maintaining required temperature in the crystallizer, and a magnetic stirrer for uniform mixing of the solution. Peristaltic pump for maintaining required antisolvent flowrate. The residual lactose concentration was determined using a refractometer (Hanna Equipment's Pvt Ltd, Mumbai, Model no HI96801). An aqueous solution of α-Lactose monohydrate was prepared and poured into the crystallizer vessel. The solution was heated 5 °C above

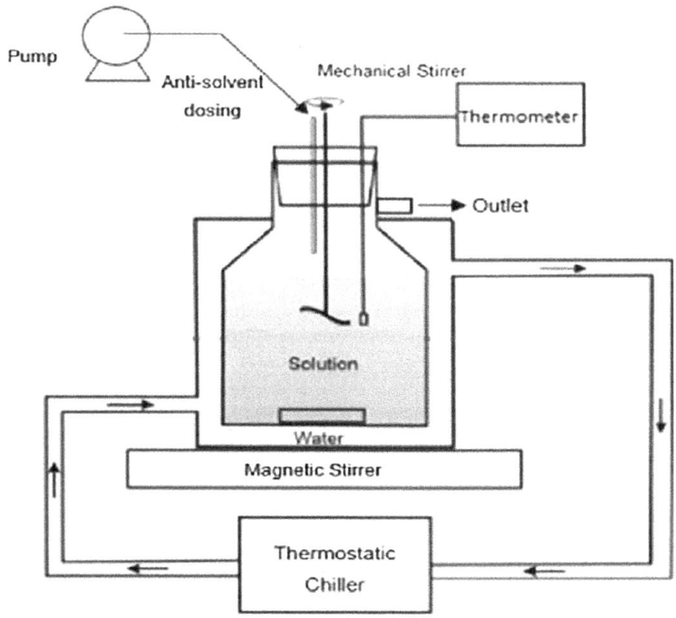

Fig. 10.1 Schematic representation of the crystallizer unit

the saturation temperature for complete dissolution of lactose particles in the solution [18]. The solution in jacketed glass crystallizer was then cooled and maintained at a constant temperature of 40 °C by thermostatic water chiller. Antisolvent addition was done at 0.1 mL/s using the peristaltic pump.

Slurry from the crystallizer was withdrawn at specific times up to 60 min. The sample was then analyzed for the lactose concentration and the average crystal size. Photographs of crystals in the slurry were taken under a microscope for image analysis. For the image analysis, ImageJ software was used. First, the conversion of scale (from pixel to micrometer) was done, followed by removal of noises and background. The watershed feature was used to separate the agglomerated particles in the image. At least 2000 crystals were counted and measured for each trial. The particle size distribution obtained from image analysis was used to calculate the mean crystal size *D* [3, 4] was determined from the number distribution obtained by image analysis by the following equation

$$D[4,3] = \frac{\sum X_i Z_i^4}{\sum X_i Z_i^3}. \tag{10.11}$$

10.4 Results and Discussion

10.4.1 Effect of Initial Concentration

The effect of initial lactose concentration (0.25, 0.30, and 0.37 kg/kg) on the crystal size and morphology was studied. The average crystal size obtained at the end of the 60 min trials for 0.25, 0.30, and 0.37 kg/kg initial lactose concentration were 26.12, 29.14, and 33.83 μm, respectively reported in Fig. 10.2a. This slight increase in initial lactose concentration was due to higher supersaturation generated leading to faster growth of crystals. The antisolvent crystallization also led to nearly 100% recovery of dissolved lactose as reported in Fig. 10.2b.

Micrographs of crystals recovered for an initial lactose concentration of 0.25, 0.30, and 0.37 kg/kg of lactose solution at 60 min are shown below in Fig. 10.3.

All photographs show tomahawk morphology and an increase in crystal number and size with time. This could be due to the slow addition rate of antisolvent allows crystals to grow only in tomahawk shape.

10.4.2 Kinetic Parameter Estimation

The data fitting was carried out for data obtained for trials 0.25, 0.3, and 0.37 kg/kg initial lactose concentration and the average values with standard deviation were reported. The simulated values of kinetic parameters using $f_{\text{minsearch}}$ for all cases are reported in Table 10.1.

The observed values of kinetic parameters showed an insignificant change in values of constants which indicates crystals followed the same kinetics with a change in initial lactose concentration. Bund and Pandit (2007) used the MSMPR equations of continuous mode for determination of crystallization kinetics for batch study and reported lower values compared to the present study [19].

10.5 Conclusions

Semi-batch antisolvent crystallization of α-lactose monohydrate was studied under varying initial lactose concentration. Results show that the size of crystals increased with an increase in the initial concentration of lactose. It was observed that almost 100% of lactose recovery was possible in 1 h. The experimental data were fitted to the model predicted values to predict the growth and nucleation kinetic parameters during the antisolvent crystallization of lactose.

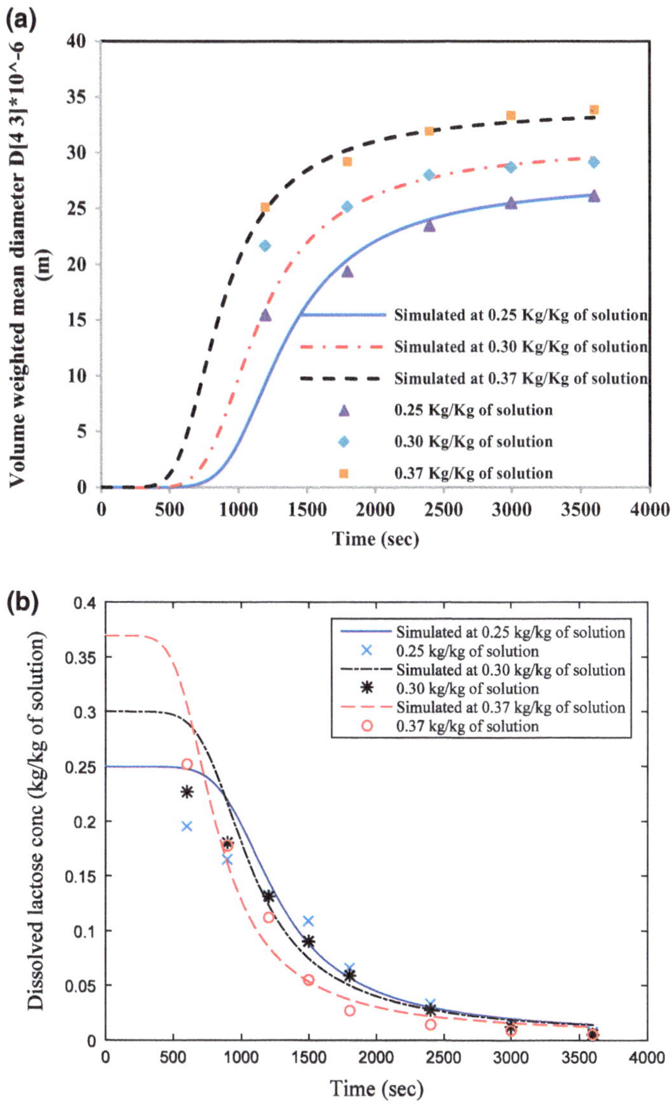

Fig. 10.2 **a** Average particle size formed with time for **a** 0.25, **b** 0.30, and **c** 0.37 kg/kg of lactose solution. **b** Residual lactose concentration profile for **a** 0.25, **b** 0.3, and **c** 0.37 kg/kg lactose solution

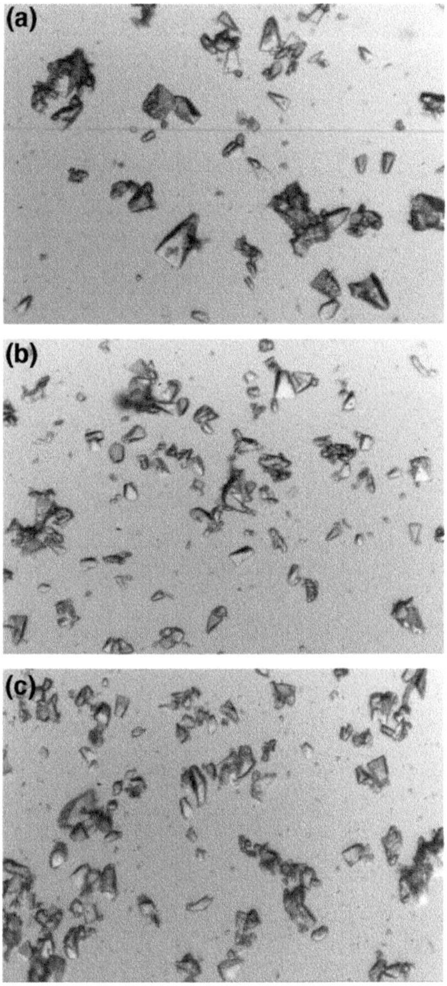

Fig. 10.3 Micrographs of crystals for **a** 0.25, **b** 0.30, and **c** 0.37 kg/kg of lactose solution. 10X magnification

Table 10.1 Crystallization kinetics constants

Kinetic constants	0.25 kg/kg	0.30 kg/kg	0.37 kg/kg	Average with std deviation
k_b (number/m^3s)	7.56E + 07	7.56E + 07	7.56E + 07	7.56E + 07
k_g (m/s)	2.29E−06	2.29E−06	2.29E−06	2.29E−06
b	0.65	0.73	0.77	0.71 ± 0.06
g	1.72	1.75	1.75	1.74 ± 0.017

References

1. Kougoulos, E., Marziano, I., Miller, P.R.: Lactose particle engineering: influence of ultrasound and anti-solvent on crystal habit and particle size. J. Crystal Growth **312**, 3509–3520 (2010)
2. Svoboda, V., Mcginty, J., Nordon, A., Sefcik, J.: Crystallization diagram for antisolvent crystallization of lactose: using design of experiments to investigate continuous mixing-induced supersaturation. Crystal Growth Design (2017)
3. Whittier, E.O.: Lactose and its utilization: a review. J. Dairy Sci. **27**, 505–537 (1944)
4. Sherwood, D.B., Raghavan, J.N.S.S.L., Ristic, R.I.: Morphology of crystals of r-lactose hydrate grown from aqueous solution. J. Phys. Chem. B **146**, 12256–12262(2000)
5. Wong, S.Y., Hartel, R.W.: Crystallization in lactose refining-a review. J. Food Sci. **79**, R257–R272 (2014)
6. Shi, T., Liang, B., Hartel, R.W.: Growth rate dispersion effects on lactose crystal size distributions from a continuous cooling crystallizer. 848–854 (1991)
7. Shi, Y., Liang, B., Hartel, R.W.: Crystallization kinetics of alpha-lactose monohydrate in a continuous cooling crystallizer. J. Food Sci. **55**, 817–820 (1990)
8. Larhrib, H., Peter, G., Marriott, C., Prime, D.: The influence of carrier and drug morphology on drug delivery from dry powder formulations. Pharmaceutical Sci. **257**, 283–296 (2003)
9. Patel, S., Murthy, Z.V.P.: Waste valorization: recovery of lactose from partially deproteinated whey by using acetone as anti-solvent. Dairy Sci. Technol. **91**, 53–63 (2011)
10. Bund, R.K., Pandit, A.B.: Sonocrystalli-zation: effect on lactose recovery and crystal habit. Ultrasonic Sonochemistry **14**, 143–152 (2007)
11. Dhumal, R.S., Biradar, S.V., Paradkar, A.R., York, P.: Ultrasound assisted engineering of lactose crystals. Pharm. Res. **25**, 2835–2844 (2008)
12. Dincer, T.D., Zisu, B., Vallet, C.G.M.R., Jayasena, V., Palmer, M., Weeks, M.: Sonocrystallisation of lactose in an aqueous system. Int. Dairy J. **35**, 43–48 (2014)
13. Yucel, U., Coupland, J.N.: Ultrasonic characterization of lactose crystallization in gelatin gels. J. Food Sci. **76**, 48–54 (2011)
14. Gajendragadkar, C.N., Gogate, P.R.: Ultrasound assisted intensified recovery of lactose from whey based on antisolvent crystallization. Ultrasonic Sonochemistry (2016)
15. Ramisetty, K.A., Pandit, A.B., Gogate, P.R.: Ultrasound-assisted antisolvent crystallization of benzoic acid: effect of process variables supported by theoretical simulations. Ind. Eng. Chem. Res. **52**, 17573–17582 (2013)
16. Nowee, S.M., Abbas, A., Romagnoli, J.A.: Antisolvent crystallization: model identification, experimental validation and dynamic simulation. Chem. Eng. Sci. **63**, 5457–5467 (2008)
17. Agrawal, S., Paterson, T., Jones, J., Mcleod, J., Bronlund, J.: Food and bioproducts processing a mathematical model based parametric sensitivity analysis of an evaporative crystallizer for lactose monohydrate. Food Bioproduct Process. **97**, 1–11 (2015)
18. Sangwal, K., Mielniczek-brz, E.: Antisolvent crystallization of aqueous ammonium dihydrogen phosphate solutions by addition of acetone at different rates. Crystal Res. Technol. **16** (2016)
19. Bund, R.K., Pandit, A.B.: Rapid lactose recovery from buffalo whey by use of "anti-solvent, ethanol". J. Food Eng. **82**, 333–341 (2007)

Chapter 11
Study on Feed and Reject Water Quality in Reverse Osmosis Systems

D. S. Bhagwat and G. R. Munavalli

Abstract Reverse osmosis (RO) is a widely used point of use treatment system for drinking water in many households. Quantity of reject water coming out of these systems is approximately 70% of feed and generally disposed off in drainages. Presently available literature lacks comprehensive knowledge regarding quality and quantity of reject coming out of household RO systems. Also, wastage of large amount of water in the form of reject makes it necessary to find a better disposal option. It can be reused for irrigation of gardens or farms, washing and flushing at the household level, etc., and can be recycled to minimize wastage. In this context, the present study is carried out to evaluate the quality and quantity of feed and reject water of various RO systems used at household. The water samples were collected and analyzed for quality parameters like pH, TDS, total hardness, Ca, Mg and MPN. The flow rate was also measured for some RO systems. Results indicated that TDS of feed water and reject water ranged between 238 to 1317 mg/L and 279 to 1616 mg/L, respectively. Sodium adsorption ratio (SAR) and residual sodium carbonate (RSC) were found to be 1.86 and −8.00, respectively. It is concluded that reject water is suitable for irrigation to garden plants and can be disposed off in soil without any treatment.

Keywords Feed · Permeate · Reject · RO · TDS

D. S. Bhagwat (✉)
Civil Environmental Engineering, Walchand College of Engineering, Sangli, Maharashtra 416415, India
e-mail: dhanashribhagwat1994@gmail.com

G. R. Munavalli
Department of Civil Engineering, Walchand College of Engineering, Sangli, Maharashtra 416415, India
e-mail: guru.munavalli@walchandsangli.ac.in

© Springer Nature Singapore Pte Ltd. 2020

V. Sivasubramanian and S. Subramanian (eds.), *Global Challenges in Energy and Environment*, Lecture Notes on Multidisciplinary Industrial Engineering, https://doi.org/10.1007/978-981-13-9213-9_11

11.1 Introduction

Clean water is one of the most important elements for all living organisms to sustain life. From the past few decades, various techniques have been developed for treating the water. Among them, the most important methods are solvent extraction, micro and ultrafiltration, sedimentation and gravity separation, flotation, precipitation, coagulation, oxidation, evaporation, distillation, reverse osmosis, adsorption, ion exchange, electrodialysis, electrolysis, etc. Membrane processes can play a key role in reducing water scarcity. Reverse osmosis (RO), nanofiltration (NF), electrodialysis reversal (EDR), ultrafiltration (UF) and microfiltration (MF) are the membrane processes which have application to drinking water [1]. Reverse osmosis is a demineralization process that relies on a semipermeable membrane to effect the separation of dissolved solids from a liquid. The semipermeable membrane allows liquid and few ions to pass, but retains the bulk of the dissolved solids (ions). The technological advancement in applications of membrane processes was efficiently explored for desalinization of seawater or brackish water in recent decades due to the huge demand for potable water. However, relatively, little improvements have been reported in the management and handling of the major by-product called reject water or brine. It is largely free from toxic components, and its composition is almost identical to that of the feed water but in a concentrated form [2]. Literature review suggests that the quantity and quality of desalinization concentrate depend on temperature, pressure of feed water, source water quality, pretreatment, desalination processes implemented and water recovery [3–6]. Various effects of application of RO reject as irrigation water for salt tolerant plants, soil and groundwater were monitored, and the issue of environmental degradation was raised [7–10]. Present study is aimed at assessing the quality and quantity of feed and reject water related to the household RO water purifier and its suitability as irrigation water for garden plants.

11.2 Methodology

The study was conducted on reverse osmosis (RO)-based point of use water purifiers which were either branded or assembled systems.

11.2.1 Sampling

Small-scale point of use water purification systems using reverse osmosis were selected based on the feed water source, feed water quality and type of use of RO system. Systems using groundwater, surface water or both as feed source were selected from Miraj, Vishrambag and Sangli. Accordingly, six systems were using groundwater, four were using surface water and one was using mixed source water as a feed source. All the systems selected were using aromatic polyamide thin film composite membranes. Feed water quality was varied from TDS 238–557 mg/L for surface water and 470–1317 mg/L for groundwater. Feed, permeate and reject samples were collected from each RO system by grab sampling method. Sampling frequency was decided once in 15 days to get variation in feed and reject water quality with respect to time and cumulative volume of permeate passed through the membrane. Accordingly, samples were taken from October 2017 to January 2018. A questionnaire was prepared and circulated among the users to collect additional information like flow rate of reject and permeate and approximate cumulative volume of permeate collected throughout the life span of the membrane, etc.

11.2.2 Water Quality and Quantity Analysis

Feed, permeate and reject samples were analyzed for pH, TDS, total hardness, ionic concentrations of Ca and Mg and MPN. Also, all the reject samples were mixed and this sample was analyzed for pH, TDS, Total Hardness, Ca and Mg concentration, sulfates, nitrates, Na, carbonates, bicarbonates, sodium adsorption ratio (SAR) and residual sodium carbonate (RSC). pH, EC and TDS were determined using multi-ion meter of Hach make. All the parameters were analyzed referring to [11].

11.3 Results and Discussions

It was observed that the quality of feed water varied from TDS 238–1317 mg/L and quality of reject water varied from 450 to 1617 mg/L for both surface and ground feed water sources.

11.3.1 Feed and Reject Quality Correlation

It was observed that reject coming out of RO systems using surface water and groundwater as feed source had pH of 6.5–7.96 and 6.47–8.01, respectively. Feed water pH was slightly higher than the reject in many cases. As in Tables 11.1 and 11.2, TDS in reject water was greater than feed water TDS by 10.5–305.5 mg/L. Total hardness of reject water was greater than feed water by 9.375–180 mg/L as $CaCO_3$. Ca and Mg concentration in reject water was greater than feed water by approximately 3.75–70 mg/L and 0.72–31.08 mg/L, respectively. Figures 11.1, 11.2, 11.3, 11.4 and Fig. 11.5 show variation in feed and reject water quality for each RO system. MPN for all reject samples was found positive.

Table 11.1 Feed water quality

Sampling station	TDS (mg/L)	Total hardness (mg/L as $CaCO_3$)	Ca Conc. (mg/L as Ca)	Mg Conc. (mg/L as Mg)
Vijaynagar	1145–1317	692–820	276.8–328	57.6–103.68
Brahmanpuri	505–558	316–468	126.4–187.2	38.4–145.92
WTP 1	470–550	352–540	140.8–216	61.44–96.96
WTP 2	597–1066	352–600	140.8–240	52.8–122.88
Pushparaj Chowk	1064–1311	368–680	147.2–272	33.6–71.04
Railway colony	503–580	240–528	96–211.2	45.12–66.24
Electrical dept.	253–537	208–408	83.2–163.2	19.2–28.8
Market yard	248–540	240–390	96–156.8	26.88–61.92
Conference room	238–529	200–400	80–160	23.04–44.16
Wanleswadi	292–1099	244–656	97.6–262.4	24.96–96.96
School	355–509	260–428	104–171.2	22.08–48

Table 11.2 Reject water quality

Samplinng station	TDS (mg/L)	Total Hardness (mg/L as $CaCO_3$)	Ca Conc. (mg/L as Ca)	Mg Conc. (mg/L as Mg)
Vijaynagar	1292–1626	816–980	326.4–392	66.24–148.8
Brahmanpuri	571–698	296–620	118.4–248	34.56–70.8
WTP 1	535–796	408–700	163.2–280	60.48–128.64
WTP 2	757–1301	424–652	169.6–260.8	53.76–135.36
Pushparaj Chowk	1155–1676	516–816	206.4–326.4	38.4–78.72
Railway colony	682–798	476–616	190.4–246.4	48–72
Electrical dept.	330–530	260–484	104–193.6	24–63.36
Market yard	222–544	216–496	86.4–198.4	25.92–76.8
Conference room	279–668	224–484	89.6–193.6	27.84–77.76
Wanleswadi	348–1190	408–880	163.2–352	38.4–222.72
School	405–558	276–504	110.4–201.6	33.6–58.56

11.3.2 Cumulative Volume of Permeate Passed Through the Membrane

Table 11.3 shows the cumulative volume of permeate passed through the RO membrane throughout its life span. It was observed that as the cumulative volume of permeate passed from a membrane throughout its life span increases, reject flow increases.

In case of sampling station 'Pushparaj Chowk,' membrane was changed after the first sampling. It was observed that reject flow was very high before changing the

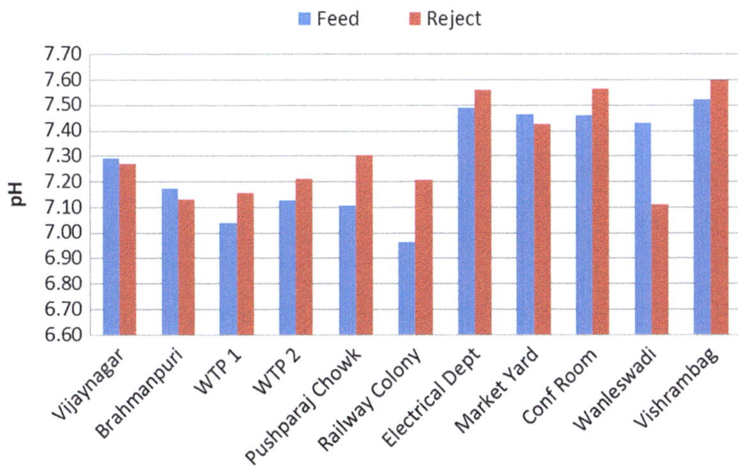

Fig. 11.1 pH

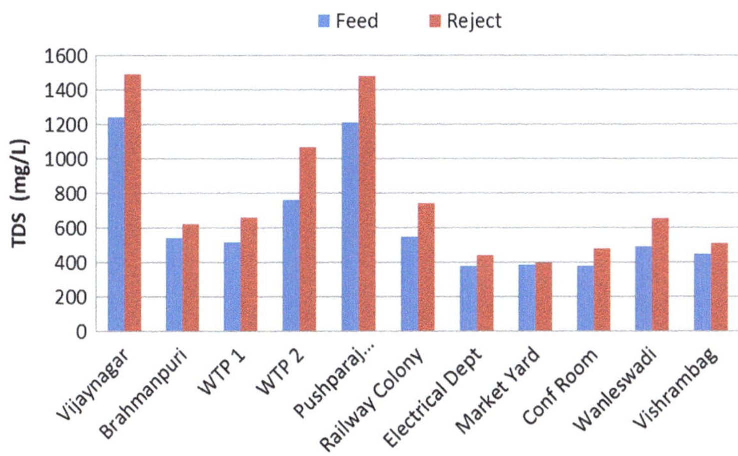

Fig. 11.2 TDS

membrane. For further samplings after changing the membrane, reject flow decreased and remained constant as shown in Fig. 11.6.

In case of sampling station 'Vijaynagar,' membrane was unchanged and the flow rate of reject increased consistently as shown in Fig. 11.7.

11.3.3 Irrigation Requirements

According to Table 11.4, it was observed that reject water TDS was 450–1617 mg/L.

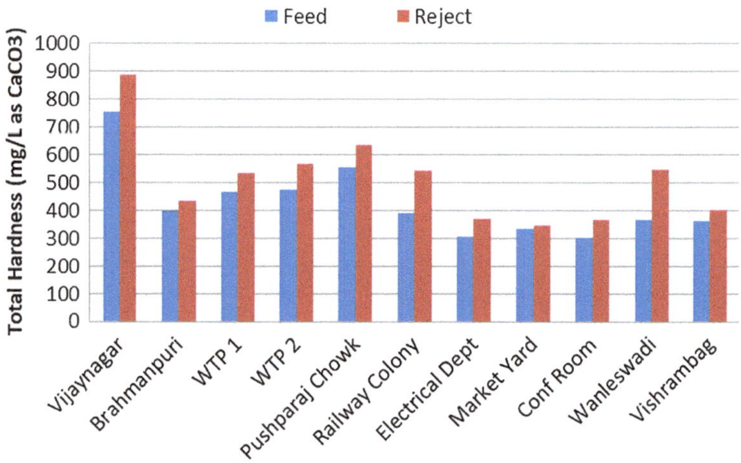

Fig. 11.3 Total hardness

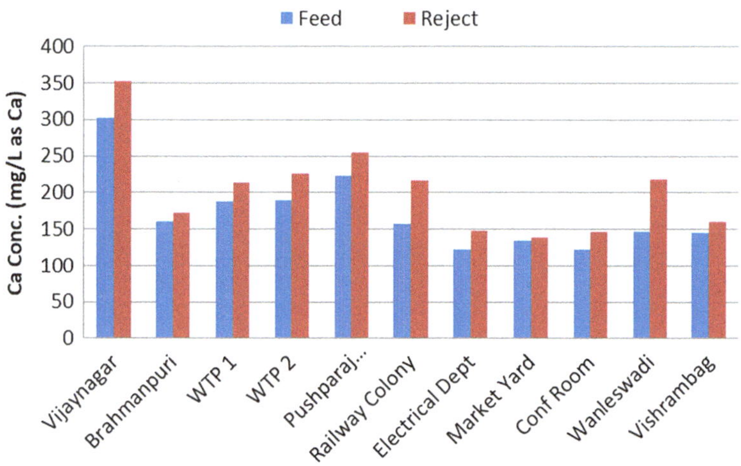

Fig. 11.4 Ca concentration

Ionic concentration of calcium is greater than ionic concentration of magnesium. Sodium adsorption ratio (SAR) potential of mixed reject sample was found to be 1.86 which is well below the permissible limit prescribed in BIS:11624 (1986). Also, (RSC) residual sodium carbonate potential was found to be −8.00 which is below 1.5.

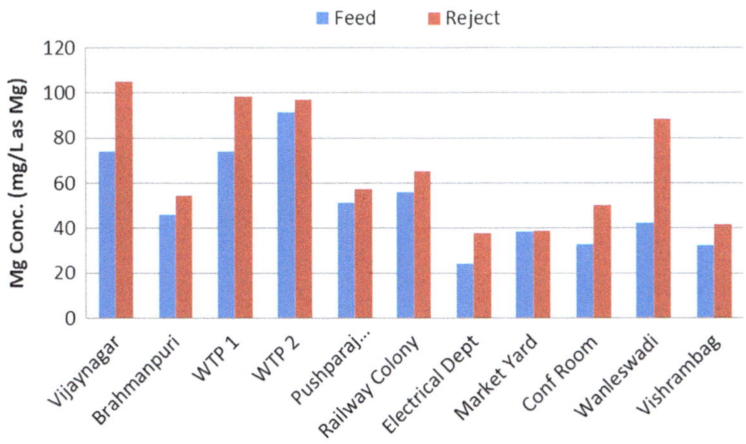

Fig. 11.5 Mg concentration

Table 11.3 Cumulative volume of permeate passed

	Month	Pushparaj Chowk	Vijaynagar
Approximate cumulative Volume of permeate Passed (Lit.)	October	10,300	1824
	November	0	1974
	December	600	2124
	January	1200	2274

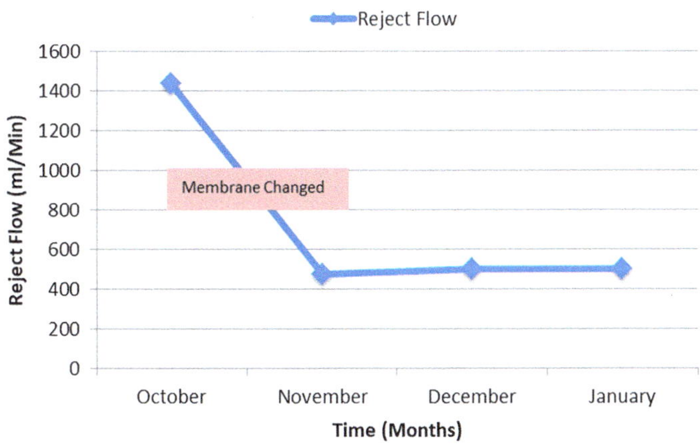

Fig. 11.6 Pushparaj Chowk—membrane changed

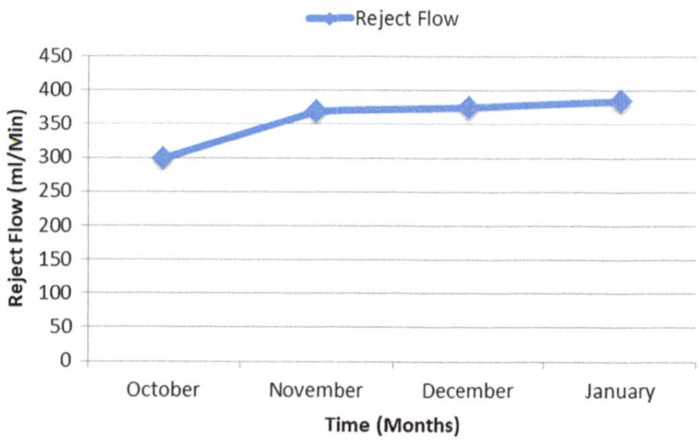

Fig. 11.7 Vijaynagar—membrane unchanged

Table 11.4 Mixed reject sample

Mixed reject sample			
Parameter	Value	Unit	Irrigation standard IS11624–1986
pH	7.52		6.5–8.5
TDS	1251	mg/l	Below 1500
Carbonates	0		
Bicarbonates	658.8	mg/l as HCO^{3-}	
Chlorides	399.876	mg/l	
Sulfates	344.56	mg/l	
Nitrates	125	mg/l	
Total Hardness	680	mg/l	
Ca concentration	272	mg/l as Ca	
Mg concentration	62.4	mg/l as Mg	
Na	24	mg/l	
SAR	1.86		Below 10
RSC	−8.00		Below 1.5

11.4 Conclusion

For small-scale household RO water purifiers, reject water quality was poor than feed water quality. As the cumulative volume of permeate passed through the membrane throughout its life span increases, the flow of reject water increases due to membrane clogging. It is concluded that the quality and quantity of reject water coming out of household water purifiers are suitable for irrigation to garden plants or hydroponics;

hence, reject water can be reused as irrigation water. Further study must be conducted on the impacts of reject water application on garden plants and soil.

Reference

1. Xu, P., Cath, T., Robertson, A., Reinhard, M., Leckie, J., Drewes, J.: Critical review of desalination concentrate management, treatment and beneficial use. Environ. Eng. Sci. **30**, 502–514 (2013)
2. Ladewig, B., Asquith B.: Desalination concentrate management, 2. Springer Briefs in Green Chemistry for Sustainability (2012)
3. Brehant, A., Bonnelye, V., Perez, M.: Assessment of ultrafiltration as a pretreatment of reverse osmosis membranes for surface seawater desalination. Water Sci. Technol. Water Supply **3**(5–6), 437–445 (2003)
4. Arora, M., Maheshwari, R., Jain, S., Gupta, A.: Use of membrane technology for potable water production. Desalination **170**, 105–112 (2004)
5. Mauguin, G., Corsin, P.: Concentrate and other waste disposals from SWRO plants: characterization and reduction of their environmental impact. Desalination **182**, 355–364 (2005)
6. Greenlee, L., Lawler, D., Freeman, B., Marrot, B., Moulin, P.: Reverse osmosis desalination: water sources, technology, and today's challenges. Water Res. **43**, 2317–2348 (2009)
7. Al-Faifi, H., Al-Omran, A., Nadeem, M., El-Eter H.,Khater, A., El-Maghraby, S.: Soil deterioration as influenced by land disposal of reject brine from Salbukh water desalination plant at Riyadh, Saudi Arabia. Desalination **250**, 479–484 (2010)
8. Soliz, D., Glenn, E., Seaman, R., Yoklic, M., Nelson, S., Brown, P.: Water consumption, irrigation efficiency and nutritional value of *Atriplex Lentiformis* grown on reverse osmosis brine in a desert irrigation district. Agr. Ecosyst. Environ. **140**, 473–483 (2011)
9. Gedam, V., Patil, J., Kagne, S., Sirsam, R., Labhasetwar, P.: Performance evaluation of polyamide reverse osmosis membrane for removal of contaminants in ground water collected from Chandrapur District. J. Membr. Sci. Technol. **2**(3) (2012)
10. Flores, A., Shukla, M., Daniel, D., Ulery, A., Schutte, B., Picchioni, G., Fernald, S.: Evapotranspiration changes with irrigation using saline ground water and RO concentrate. J. Arid Environ. **131**, 35–45 (2016)
11. APHA.: Standard method for examination of water and wastewater, 22nd edn. American Public Health Association, Washington DC (2012)

Chapter 12
Synthesis of TiO_2 Incorporated Polycaprolactone Based Ultrafiltration Membranes for Water Treatment

Jose Peter, Jobin Jose, Amal Thomas, Jefin C. Joy, S. Nivedita and Shiny Joseph

Abstract This study focused on the synthesis of polycaprolactone based ultrafiltration membranes incorporated with titanium (IV) oxide (TiO_2) nanoparticles using the phase inversion technique. The membrane was optimized at 18 wt% PCL before the incorporation of TiO_2. The study covered a range of compositions by the addition of TiO_2 from 0.5 to 2 wt%. The synthesized membranes were then tested for properties such as porosity, contact angle, mechanical strength and then characterized using TGA and FTIR techniques to determine the functional groups present and to analyze the thermal stability. Further, SEM images were recorded to analyze the pore morphology. Finally, the optimized membrane was tested to determine the pure water flux.

Keywords Hydrophilicity · Phase inversion · Polycaprolactone · TiO_2 · Ultrafiltration

12.1 Introduction

Water is one of the most important natural resources in which only a small fraction of its total quantity is available to use. This is due to the disposal of pollutants like heavy metals, dyes, pharmaceuticals, pesticides, fluorides, phenols, insecticides, pesticides, and detergents, which is a matter of serious concern that requires novel purification techniques [1].

In view of this, the development of cost-effective and robust methods for wastewater treatment requires greater attention than never before. These treatment techniques should be so devised such that they do not further stress the environment or endanger human health [2]. Although a number of methods such as adsorption, coagulation, membrane process, photocatalytic degradation, and biological methods have been used for purifying wastewater, their applications have been often undermined by

J. Peter · J. Jose · A. Thomas · J. C. Joy · S. Nivedita · S. Joseph (✉)
Department of Chemical Engineering, National Institute of Technology Calicut, Calicut 673601, Kerala, India
e-mail: shiny@nitc.ac.in

© Springer Nature Singapore Pte Ltd. 2020

V. Sivasubramanian and S. Subramanian (eds.), *Global Challenges in Energy and Environment*, Lecture Notes on Multidisciplinary Industrial Engineering, https://doi.org/10.1007/978-981-13-9213-9_12

factors, such as energy requirement, processing efficiency, economic benefit, and infrastructure [3].

Membrane filtration is one of the treatment techniques widely used for water purification projects. This process is capable of removing suspended solids and organic compounds as well as inorganic contaminants such as heavy metals. One of its main advantages is that it consumes less energy. Depending on the size of the particle that can be retained, various types of membrane filtration such as ultrafiltration, nanofiltration, and reverse osmosis can be employed. Ultrafiltration operates on separating compounds of 1000–100,000 Da using pore sizes of 10–100 nm. Polymer-supported ultrafiltration (PSU) technique has been shown a promising alternative for the removal of pollutants from industrial effluents.

Polycaprolactone, prepared by ring-opening polymerization of caprolactone, is a hydrophobic, semicrystalline polymer. Its exceptional blend compatibility has contributed toward research into its potential application in the biomedical field and it has been found safe to human health [4, 5]. PCL has a glass transition temperature (T_g) of −60 °C and a melting point ranging between 59 and 64 °C. It is soluble in various solvents such as chloroform, dichloromethane, carbon tetrachloride, benzene, toluene, cyclohexanone, and 2-nitropropane at room temperature. The main advantage of PCL over conventionally used membrane materials is that it is very much cheaper and also is a biodegradable polymer.

TiO_2 particles have the property to adsorb heavy metals efficiently and thus enhance the separation [6]. TiO_2 is also known to have the self-cleaning ability because of its antibacterial properties [7]. Presence of TiO_2 reduces the contact angle which is a direct measurement of hydrophilicity, requiring only small amounts of TiO_2 to get significant improvements [8]. TiO_2 is a pore-forming agent as its presence will increase water diffusion into the growing membrane during phase inversion process and is also known to have resistance to the fouling problem through its enhanced hydrophilicity [9].

Phase inversion is used to prepare membranes of a variety of morphologies ranging from very porous microfiltration membranes to more dense reverse osmosis membranes. It involves the controlled transformation of a polymer in a solution state to a solid state and includes a broad array of different techniques such as immersion precipitation, thermal-induced phase separation, vapor-phase precipitation, and precipitation by controlled evaporation. Out of these, immersion precipitation is the most widely used technique for commercial production of membranes where a polymer solution is cast on a suitable support. Later, it undergoes precipitation induced by the exchange of solvent and non-solvent in a coagulation bath of non-solvent giving the final membrane structure, which is a result of the combination of mass transfer and phase separation [10].

This work focuses on the preparation of TiO_2 incorporated PCL based ultrafiltration membrane and to study the effect of weight percentage of TiO_2 on the membrane characteristics and to determine the pure water flux of the optimized membrane.

12.2 Experiments

12.2.1 Materials

Polycaprolactone (PCL) pellets with a molecular weight of Mw = 80,000 were purchased from Sigma-Aldrich for membrane fabrication. N, N-Dimethylformamide (DMF) was purchased from Sisco Research Laboratories and used as a solvent. Titanium (IV) oxide nanoparticles, a mixture of rutile and anatase 33–37 wt% in H_2O dispersion, purchased from Sigma-Aldrich were used to study its effect on membrane characteristics. Deionized (DI) water was used as the non-solvent in the phase inversion immersion precipitation. Other chemicals were purchased from local suppliers and used as received.

12.2.2 Membrane Synthesis

PCL membranes were synthesized by phase inversion immersion precipitation method. The calculated amount of PCL was taken in a beaker along with 10 mL of DMF solvent. The mixture was stirred at a rpm of about 150–180 at a temperature of 65 °C for a period of about 1.5 h until the pellets melted and uniformly dissolved in the solvent. Then the required weight of TiO_2 nanoparticles was added into the polymer solution under vigorous stirring. When the resulting solution became homogeneous and viscous, the stirring was stopped and was cast on a clean glass plate at room temperature using a casting blade set to give a film of thickness 200 μm. After about 2 min, the glass plate along with the polymer film was immersed gently in a coagulation bath of DI water (non-solvent), so that phase inversion occurs and the film solidifies. Later, the solidified film was peeled off and left in the water bath for 24 h containing 2–3 drops of formaldehyde solution to prevent bacterial growth. Later, the membrane was dried at room temperature for another day. The compositions of the prepared membranes are presented in Table 12.1.

Table 12.1 Compositions of the synthesized membranes

Membrane	Membrane compositions		
	PCL (wt%)	TiO_2 (wt%)	Volume of DMF (mL)
D1	18	0	10
T1	17.5	0.5	10
T2	17	1	10
T3	16.5	1.5	10
T4	16	2	10

12.2.3 Membrane Characterization

12.2.3.1 Membrane Porosity

Samples of a specific area (*S*) were cut off from the synthesized membranes and immersed in a DI water bath for 24 h. The weight of the immersed samples (W_w) in grams and their wet thickness (l) in cm were calculated after wiping them off using tissue paper. The thickness was calculated using a digital micrometer. Later, the samples were left to dry in an oven at 50 °C for 24 h and the dry weight was calculated (W_d) in grams. The porosity (Pr) was then calculated in percentage using Eq. (12.1):

$$P_r = \frac{W_w - W_d}{Sl\rho} \tag{12.1}$$

where ρ (g/cc) is the density of water.

12.2.3.2 Hydrophilicity

Hydrophilicity of the membranes was estimated by the calculation of the contact angles using an automated contact angle goniometer using DI water as the liquid. Dry membrane samples of dimensions of about 4×0.5 cm^2 were cut out for the test. The liquid dosing was software-controlled and a motor-driven microsyringe placed the water droplet on the dry membrane surface. The average value of the contact angle was estimated based on the readings taken at about four arbitrary points.

12.2.3.3 Mechanical Strength

Mechanical strength is an important property required by ultrafiltration membranes since they are usually operated under high pressures ranging 1–10 bar [11]. The tensile strength of the membranes was calculated using the Shimadzu AG-X plus 10 kN universal testing machine (UTM) in accordance with the ASTM D882 standards. They were vertically mounted between the two gripping units of the tester. A load of 9 kN was employed at a crosshead speed of 5 mm/min. The average value was calculated by using about three samples for each membrane composition.

12.2.3.4 Scanning Electron Microscopy (SEM)

Morphology and pore distribution were observed using SEM imaging technique. The top surface of D1 and T3 membranes was analyzed using Hitachi SU-6600 FESEM to infer the effect of TiO_2 on the membrane surface.

12.2.3.5 Thermo Gravimetric Analysis (TGA)

Weight degradation of PCL membranes was determined by TGA conducted on the Hitachi STA7200. Sample weighing 10 mg of D1 and T3 membranes was heated in nitrogen atmosphere from 25 to 500 °C at a rate of 10 °C/min.

12.2.3.6 Fourier-Transform Infrared Spectroscopy (FTIR)

Functional groups present in the synthesized membranes were identified by FTIR conducted on a 630 1B diamond ATR module in the range of wavenumbers 4000–400 cm^{-1}. The test was conducted on samples of D1 and T3 membranes.

12.2.3.7 Water Flux Measurement

Water flux through the T3 membrane was measured using a cross-flow filtration unit to identify the permeability of the membranes. The test was conducted on a membrane sample having an effective membrane area of 42 cm^2. After compacting the samples at a pressure of 12 bars for 30 min, the steady-state fluxes were measured at a pressure of 4 bars using Eq. (12.2):

$$J_w = \frac{Q}{A\Delta t} \quad (12.2)$$

where J_w (L/m^2h) is the water flux, Q (L) is the volume of water that permeated, A (m^2) is the effective area of the membrane used, and Δt (h) is the time elapsed.

12.3 Results and Discussion

12.3.1 Porosity

Table 12.2 shows the porosity and contact angle of the synthesized membranes. It can be seen that the porosity increases initially with an increase in TiO_2 concentration. Comparable values were obtained for membranes T2 and T3. This is followed by a drop because of the formation of a more compact membrane structure due to the high viscosity of the polymer solution and higher density of polymer chains [12].

Table 12.2 Porosity and contact angle of the synthesized membranes

Membrane	Porosity (%)	Contact angle (°)
D1	22.18	87.60
T1	35.59	75.60
T2	64.02	67.05
T3	65.54	65.10
T4	40.14	62.20

12.3.2 Contact Angle

The wettability or hydrophilicity of a membrane is quantified using the value of the contact angle. A higher contact angle corresponds to hydrophobicity and PCL has been found to be a hydrophobic polymer [13]. It was observed that adding TiO_2 nanoparticles helps to improve the hydrophilicity of the membrane. Large drop in contact angle was observed in the case of initial TiO_2 additions. At higher TiO_2 concentrations, even though the extent of the decrease in contact angle was less, the values decreased nevertheless. This decrease is in accordance with the previous studies and has found that TiO_2 is capable of imparting hydrophilicity, in turn, contributing to antifouling of membrane surfaces.

12.3.3 Mechanical Strength

Tensile strength of the synthesized membranes has been plotted in Fig. 12.1. It is seen that the addition of the nanoparticle additive increases the tensile strength of the samples till T3 composition. This was attributed to the macrovoid suppression due to the introduction of nanoparticles in the membrane matrix [14]. However, further addition of TiO_2 greatly undermines the mechanical strength. This sudden drop is attributed to nonuniform stress distribution within the membrane after a critical TiO_2 loading when subjected to a load. Mechanical strength is an important property required by ultrafiltration membranes owing to their high operating pressure and the extent of improvement seen in T3 is appreciable.

12.3.4 Water Flux Measurement

Water flux measurements were carried out for T3 membrane and the results have been plotted in Fig. 12.2. The fluxes were measured at intervals of 10 min at a pressure of 4 bars and were found to decrease until a steady state was reached. The flux initially declined and then reached a steady state after a period of 40 min. Steady-state fluxes were also measured across the T3 membrane at different operating pressures between

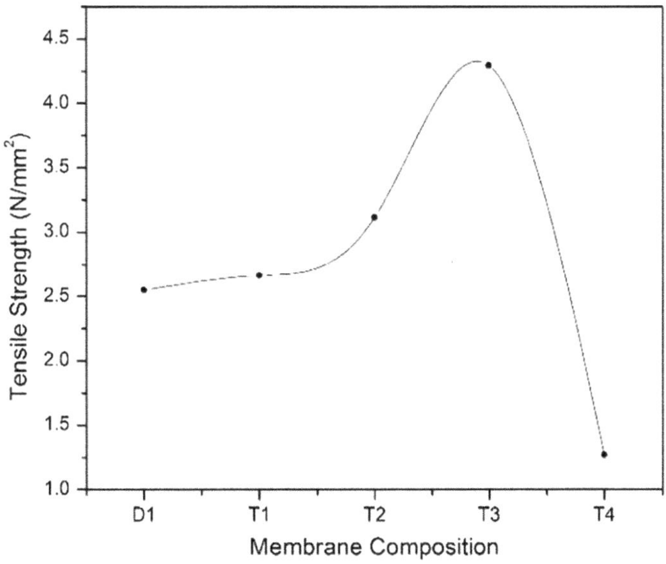

Fig. 12.1 Tensile strength of synthesized membranes

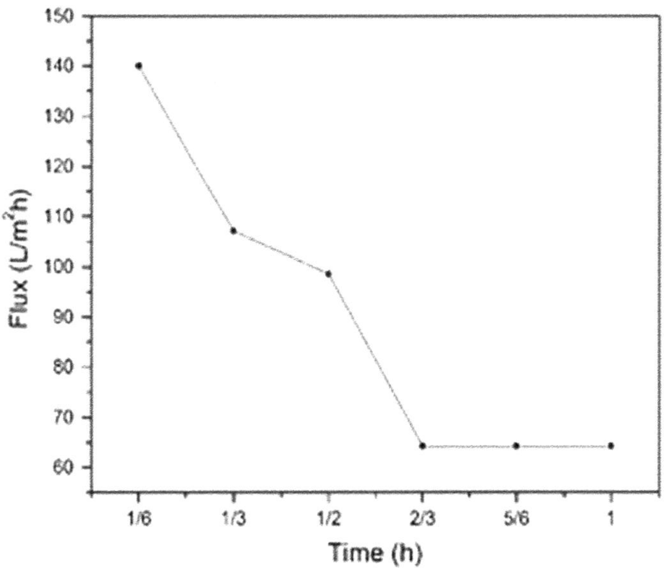

Fig. 12.2 Flux across T3 membrane versus time at 4 bars

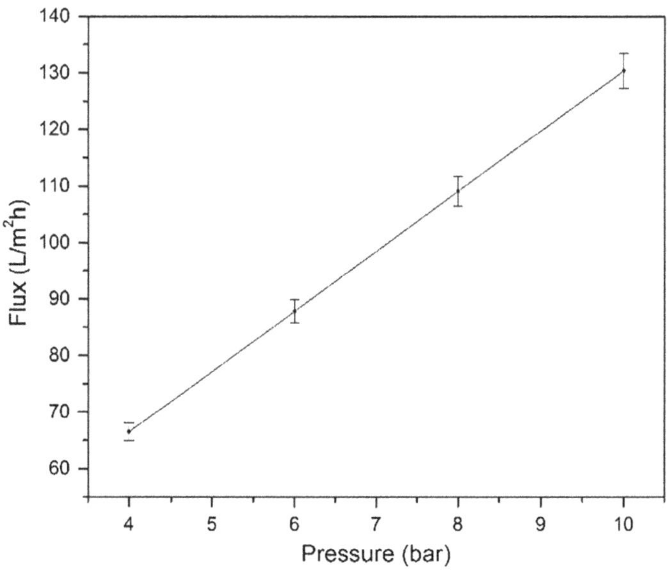

Fig. 12.3 Flux across T3 membrane versus pressure

4 and 10 bars. The fluxes were found to increase with the increase in pressure as shown in Fig. 12.3.

12.3.5 Pore Radius Calculation

The Guerout–Elford–Ferry equation was used to calculate the average pore radius (r_m) of the T3 membrane. The equation is as shown:

$$r_m = \sqrt{\frac{(2.9 - 1.75\varepsilon) \times 8l\eta Q}{\varepsilon \times \Delta P \times A}} \tag{12.3}$$

where r_m gives the mean pore radius (m), ε is the porosity of the membrane, l is the thickness of the membrane (m), η is the viscosity of water (Pa.s), and Q is the volume of water that permeates through the membrane in unit time (m^3/s). ΔP is the operational pressure (Pa) and A is the effective membrane area (m^2) [15]. Using Eq. (12.3), the mean pore radius was found to be 7.32 nm. This indicates that the pore diameter, which is double of the calculated value, lies within the range of ultrafiltration pore diameter of about 10–100 nm.

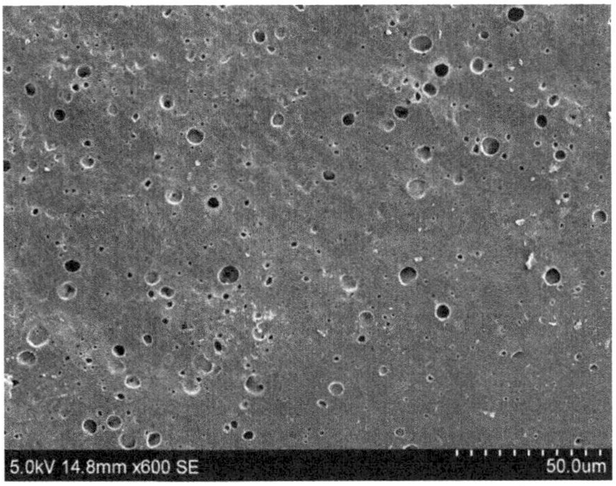

Fig. 12.4 SEM image of D1 membrane

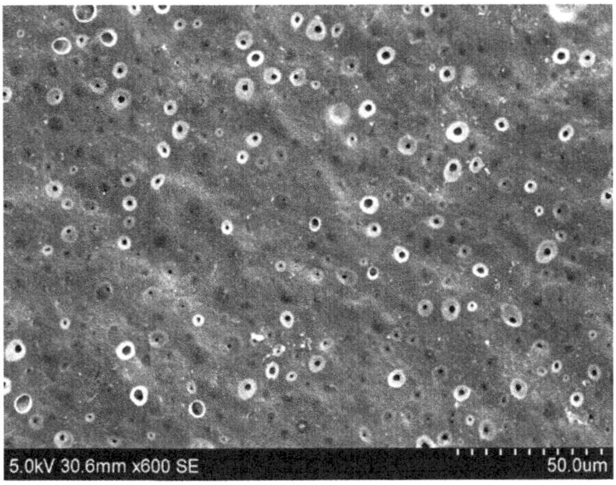

Fig. 12.5 SEM image of T3 membrane

12.3.6 SEM Analysis

SEM images of the D1 and T3 membranes are shown in Figs. 12.4 and 12.5, respectively. It was observed that the surfaces of both the membranes have got pores mostly in a circular shape. From the figures, we see that the T3 membrane has got a more uniform pore distribution than the D1 membrane.

12.3.7 TGA

Figure 12.6 shows the TGA plot of the D1 and T3 membrane samples. It was observed that the thermal decomposition of the D1 membrane starts at about 420 °C. TGA is

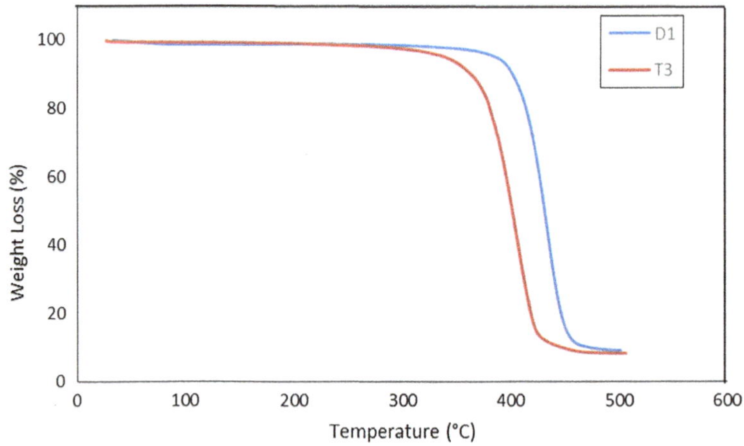

Fig. 12.6 TG curve of D1 and T3 membranes

used to analyze thermal stability and 420 °C corresponds to the temperature at which the single weight loss step was observed for the sample [13, 16]. Upon incorporating TiO_2 nanoparticles, thermal degradation starts at 360 °C. But as before a single weight loss step was observed again. Moreover, degradation of T3 was similar to that of D1 with no new residue being left behind as is evident from the converging final points of the TG plot.

12.3.8 FTIR Analysis

The FTIR spectra of D1 and T3 membranes have been plotted in Fig. 12.7. The band at 3444 cm^{-1} and the very sharp signal at 1722 cm^{-1} are attributed to the hydroxyl

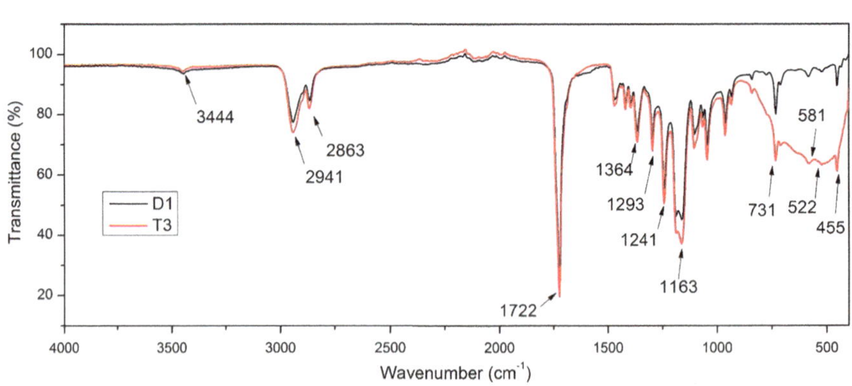

Fig. 12.7 FTIR spectra of D1 and T3 membrane

and ester groups, respectively [17]. The peak observed at 2945 cm^{-1} was found to be due to the stretching of the asymmetric $-CH_2$ bond while symmetric stretching of $-CH_2$ bonds was responsible for the one seen at 2863 cm^{-1} [18, 19]. The −COO bond stretching was found to be responsible for the peak observed at 1364 cm^{-1} while the symmetric C–O–C stretching vibrations in the repeated $-OCH_2CH_2$ unit gave rise to a peak at 1163 cm^{-1} [20]. The peak at 1293 cm^{-1} was credited to the C–C backbone and C–O stretching modes in the crystalline PCL. Asymmetric C–O–C stretching gave rise to a peak at 1241 cm^{-1} [19]. The FTIR image of T3 membrane closely resembles D1 membrane. Range of broad peaks observed between 455 and 731 cm^{-1} is attributed to the vibration absorption of TiO_2 [21].

12.4 Conclusion

In the course of this study, a variety of membranes differing in terms of the amount of TiO_2 incorporated were synthesized. Data were collected by conducting various tests and based on the results an optimum concentration was chosen. From Table 12.1, we see that T1 membranes lack in porosity, hydrophilicity, and tensile strength. Moreover, the higher contact angle indicates hydrophobicity that may in the future lead to fouling of the membrane. At the other end, T4 membranes possess better hydrophilicity but lack in porosity and tensile strength. The intermediate compositions of T2 and T3 thus provide better alternatives and among these it was found that the T3 membranes have higher tensile strength, porosity, and better hydrophilicity, both of which are properties essential for ultrafiltration membranes. Hence, the membrane incorporated with 1.5 wt% TiO_2 was optimized and the steady state water flux at pressures ranging from 4 to 10 bars was measured and analyzed. Also characterization of the membrane was conducted through TGA, FTIR, and SEM.

References

1. Schwarzenbach, R.P., Escher, B.I., Fenner, K., Hofstetter, T.B., Johnson, C.A., et al.: The challenge of micropollutants in aquatic systems. Science **313**, 1072–1077 (2006)
2. Shannon, M.A., Bohn, P.W., Elimelech, M., Georgiadis, J.G., et al.: Science and technology for water purification in the coming decades. Nature **452**, 301–310 (2008)
3. Pontius, F.W.: Water Quality and Treatment, 4th edn. US, McGraw-Hill, New York (1990)
4. Chandra, R., Rustgi, R.: Biodegradable polymers. Progr. Polym. Sci **23**, 1273–1335 (1998)
5. Nair, L.S., Laurencin, C.T.: Biodegradable polymers as biomaterials. Progr. Polym. Sci **32**, 762–798 (2007)
6. Mohamed, E.M., et al.: Solid-solid crosslinking of carboxymethyl cellulose nanolayer on titanium oxide nanoparticles as a novel biocomposite for efficient removal of toxic heavy metals from water. Int. J. Biol. Macromol. **105**, 1269–1278 (2017)
7. Adawiya, J.H., et al.: Exploring potential environmental applications of TiO_2 nanoparticles. Energy Procedia **119**, 332–345 (2017)

8. Nevstrueva, D., et al.: Effect of a TiO_2 additive on the morphology and permeability of cellulose ultrafiltration membranes prepared via immersion precipitation with ionic liquid as a solvent. Cellulose **22**, 3865–3876 (2015)
9. Sotto, A., et al.: Nanofiltration membranes enhanced with TiO_2 nanoparticles: a comprehensive study. Desalination Water Treatment **34**, 179–183 (2011)
10. Mulder, M.: Phase inversion membranes, Membrane preparation, University of Twente, Enschede, pp. 3331–3346. Academic Press, The Netherlands (2000)
11. Nath, K.: Membrane Separation Processes, 2nd edn. New Delhi, India, PHI Learning (2011)
12. Mbareck, C., et al.: Elaboration, characterization and application of polysulfone and polyacrylic acid blends as ultrafiltration membranes for removal of some heavy metals from water. J. Hazard. Mater. **171**, 93–101 (2009)
13. Morouço, P., Biscaia, S., et.al.: Fabrication of poly(ε-caprolactone) scaffolds reinforced with cellulose nanofibers, with and without the addition of hydroxyapatite nanoparticles. BioMed. Res. Int. 1–10 (2016)
14. Demirel, E., et al.: Fe_2O_3 nanocomposite PVC membrane with enhanced properties and separation performance. J. Membr. Sci. **529**, 170–184 (2017)
15. Rezaee, R., et al.: Fabrication and char-acterization of a Polysulfone-graphene oxide nanocomposite membrane for arsenate rejection from water. J. Environ. Health Science Eng. **13**(61), 1–11 (2015)
16. Xiong, R., Hameed, Nishar, Guo, Qipeng: Cellulose/polycaprolactone blends regenerated from ionic liquid 1-butyl-3-methylimidazolium chloride. Carbohyd. Polym. **90**, 575–582 (2012)
17. Benkaddour, A., Jradi, K., Robert, S., Daneault, C.: Grafting of polycaprolactone on oxidized nanocelluloses by click chemistry. Nanomaterials **3**, 141–157 (2013)
18. Safaeijavan, R., Soleimani, M., et al.: Biological behavior study of gelatin coated PCL nanofiberous electrospun scaffolds using fibroblasts. J. Paramedical Sci. **5**, 67–73 (2014)
19. Elzein, T., Nasser-Eddine, M., et al.: FTIR study of polycaprolactone chain organization at interfaces. J. Colloid Interface Sci. **273**, 381–387 (2004)
20. Abderrahim, B., Abderrahman, Ei, et al.: Kinetic thermal degradation of cellulose, polybutylene succinate and a green composite: comparative study. World J. Environ. Eng. **3**, 95–110 (2015)
21. Shoja, M., et al.: Preparation and characterization of Poly (Ɛ-caprolactone)/TiO_2 microcomposites. Digest J. Nanomater. Biostructures **10**, 471–477 (2015)

Chapter 13
Treatment of Dairy Wastewater Using the Fenton's Oxidation Process (FOP)

Anand S. Jaltade, Arvind M. Mokadam and Mangesh L. Gulhane

Abstract Nowadays, there is an increased concern over the treatment of industrial effluents. Advanced oxidation processes (AOPs) can be effective due to the recalcitrant and refractory nature of complex pollutants. AOPs provide a faster means of organic pollutant removal and thus may prove instrumental in reducing the treatment time of wastewater. Dairy wastewater normally has high organic content. This paper includes the study of the effect of Fenton's oxidation process (FOP) on the degradation of dairy wastewater. Fenton's oxidation process uses hydrogen peroxide and ferrous sulfate to form a highly oxidizing species called as hydroxyl-free radicals (·OH) which can be used to degrade the pollutants in wastewater. Especially, parameters like initial and final turbidity, pH, odor, COD and the ratio of Fenton's reagent dose to the volume of sample are considered for the study.

Keywords Fenton's process · Advanced oxidation · Dairy wastewater · Hydroxyl radical · Hydrogen peroxide · Ferrous ions

13.1 Introduction

Industrial wastewater treatment is gaining importance with the increase in the complexity of pollutants. The use of new technologies and processes for production of commodities and goods is leading to the formation of such pollutants which are more refractory to the conventional systems of treatment. Use of advanced oxidation processes (AOPs) can be seen as a prospective solution to the problem of refractory nature of modern pollutants. There are numerous AOPs available for a variety of wastewaters generated in the society viz., Fenton's oxidation, ozonation, photocatalysis, cavitation, etc. Majority of these processes involves the use of a strong oxidizing species known as hydroxyl-free radical (·OH). Among AOPs, two

A. S. Jaltade (✉)
Government College of Engineering, Amravati 444604, India
e-mail: asjcivil2015@gmail.com

A. M. Mokadam · M. L. Gulhane
Department of Civil Engineering, Government College of Engineering, Amravati 444604, India

© Springer Nature Singapore Pte Ltd. 2020

V. Sivasubramanian and S. Subramanian (eds.), *Global Challenges in Energy and Environment*, Lecture Notes on Multidisciplinary Industrial Engineering, https://doi.org/10.1007/978-981-13-9213-9_13

of the most important processes of hydroxyl radical generation are the Fenton and photo-Fenton (Fe^{2+}/H_2O_2 and $Fe^{2+}/H_2O_2/UV$) systems. Fenton's reagent, a mixture of ferrous (Fe^{2+}) ions and hydrogen peroxide that produces hydroxyl radicals, has been extensively used to oxidize organic matter in water, and to reduce the chemical oxygen demand (COD) [1]. FOP is characterized by the formation of hydroxyl-free radical when ferrous sulfate is combined with hydrogen peroxide in an aqueous solution. The reactions are mentioned below:

$$Fe^{2+} + H_2O_2 \rightarrow Fe^{3+} + OH^- + \cdot OH \tag{13..1}$$

$$Fe^{3+} + H_2O_2 \rightarrow Fe^{2+} + HO_2\cdot + H^+ \tag{13.2}$$

Gogate et al. stated that the pH of the system has been observed to significantly affect the degradation of pollutants. The optimum pH has been observed to be 3.0 in the majority of the cases and hence is recommended as the operating pH. At lower pH ($pH < 2.5$), the formation of $(Fe(II)\ (H_2O))^{2+}$ occurs, which reacts more slowly with hydrogen peroxide and therefore, produces less amount of ·OH radicals thereby reducing the degradation efficiency. Also, at operating pH of greater than 4.0, the decomposition rate decreases because of the decrease of the free iron species in the solution, probably due to the formation of Fe^{2+} complexes with the buffer inhibiting the formation of free radicals and also due to the precipitation of ferric oxy-hydroxides (FeO(OH)) which inhibit the regeneration of ferrous ions. Also, the oxidation potential of ·OH radical is known to decrease with an increase in the pH [2].

It is also seen that the degradation rate declines by dwindling the ferrous ion's concentration. Similarly, excess concentration of ferrous ions can increase the unused ferrous salt in the solution thus increasing the total dissolved solids in the solution which is not desirable. Increase in the initial concentration of hydrogen peroxide has a positive effect on the degradation process. However, its excess concentration also may contribute to the increase in the COD of the solution. High initial concentration of the pollutants has a staggering effect on the degradation efficacy of the Fenton process. The evaluation of the Fenton process is generally based on the comparison between influent and effluent parameters of the wastewater like turbidity, COD, total suspended solids and pH.

Normally, dairy wastewater is characterized by high water requirements, and the dairy industry generates wastewater with high organic content. Statistically, for every liter of milk generated and processed, 2.5 L of water is required [3]. According to the statistics of the National Dairy Development Board of India, 155.5 million tonnes of milk were processed in India in the year 2015–16 alone. Thus, the dairy industry can be considered to be a substantial source of industrial wastewater which needs treatment. When fresh, the dairy effluent is slightly alkaline and due to rapid fermentation of the milk sugar into lactic acid, it turns acidic. It has high suspended solids as well as dissolved solids concentration. The presence of these organics is the reason behind the high BOD and COD of the dairy effluent. Dairy effluents decompose rapidly and deplete the dissolved oxygen level of the receiving streams immediately resulting in anaerobic conditions and release of strong foul odors due to nuisance conditions. [4]. It is also known that the dairy effluents are responsible for the growth of sewage fungi which are found at the bottom of the streams. Thus, it becomes imperative to treat the dairy wastewater effectively to reduce its pollution potential. In the present study, effect of the various doses of Fenton reagent on the dairy wastewater as well as the synthetic dairy effluent samples is analyzed. The wastewater characteristics that are considered for the analysis are pH, turbidity, TSS and COD concentration.

13.2 Experimental Procedures

13.2.1 Apparatus and Equipment

Use of batch method was made to conduct the experiments. Glass beakers of 250 and 500 ml capacity were used as reactors for the Fenton process. Digital pH meter (SensION$^+$ PH3 by HACH) was used for measuring the pH of the solution during the reaction. For turbidity measurement, digital turbidimeter (2100Q by HACH) was used. This instrument has a range of 0–1000 NTU. TSS concentration was analyzed by using digital colorimeter (DR/890 by HACH). This instrument has a range of 0–1067 mg/l. COD was calculated by using standard open refluxing titrimetric method. All the apparatuses were washed in distilled water and dried before use.

13.2.2 Chemicals

Dairy wastewater sample was collected from the Government Milk Processing Unit stationed at Akola, Dist. Akola, as well as synthesized in the laboratory. This processing unit performs pasteurization, skimming and homogenization of an average 50,000–60,000 L of milk per day. Wastewater sample was obtained after cleaning of the entire plant at a shift's end.

For the Fenton process, hydrogen peroxide (H_2O_2, 30% w/w) (Fisher Scientific) was used. Ferrous sulfate heptahydrate ($FeSO_4$. $7H_2O$) (Fisher Scientific) was used. Standard solutions of potassium dichromate ($K_2Cr_2O_7$), ferrous ammonium sulfate (Mohr's Salt) ($(NH_4)_2FeSO_4.6H_2O$) and concentrated sulfuric acid (H_2SO_4), ferroin indicator solution were used for the COD measurement process.

13.2.3 *Procedure*

13.2.3.1 Turbidity Measurement

Samples in increasing volumes from 50 to 300 ml were taken in glass beakers at room temperature for finding the turbidity removal. The initial pH was recorded. Fenton reagent was added to the reactor beakers in two stages. First, there was addition of ferrous sulfate to the beakers. The contents were stirred with the help of a mechanical stirrer for 1 min to ensure sufficient mixing. The Fenton reaction was initiated by the second stage when hydrogen peroxide was added to the reactor beakers. The ratio of ferrous sulfate to hydrogen peroxide was 1:3 (by weight) for all the dosages. The contents of the reactor were given a stir again for 1 min at the start of the reaction. The change in the pH was recorded over a total time of 60 min at the interval of every 5 min. Then after a detention time of 1 h, the sample was decanted and used for the analysis.

13.2.3.2 Total Suspended Solids Measurement

For the measurement of TSS concentration, sample volume adopted was 200 ml. Initial turbidity and pH were recorded for the sample. Synthesized sample was used for analysis. Fenton reagent was added in similar manner as explained in Sect. 13.2.3.1. The contents were given a stir for 1 min, and a treatment time of 60 min was considered. After the treatment time, the sample was detained for 1 h, decanted and used for analysis.

13.2.3.3 COD Measurement

For the open reflux method, 20 ml volume of distilled water and the samples after treatment were taken into the COD flasks. Serial dilution with a dilution factor of 20 was adopted for the synthesized dairy wastewater sample. Mercuric sulfate (400 mg), potassium dichromate (0.10 N, 10 ml), concentrated sulfuric acid (30 ml) and silver sulfate (200 mg) were added to the flasks, and then the assembly was kept in the COD digestor at 150 °C $\pm$ 5 °C for 2 h. After digestion, the solution was added with around 90 ml of distilled water and allowed to cool to room temperature. After cooling, the solution was transferred to conical flask, and 2–3 drops of ferroin indicator were added. Ferrous ammonium sulfate (0.25 N) was standardized every time it was used as a titrant. The titration was done till the sharp color change from blue–green to wine red was observed in the solution. The change in the pH, odor, turbidity, TSS concentration and COD was recorded with respect to the initial values, and analysis was carried out based on these observed changes.

13.3 Results and Discussion

The results obtained for the given study are mentioned in the Table 13.1 below.

It can be seen from Table 13.1 that the reduction in COD in case of dairy wastewater is 52.04%. Similarly, for the synthesized sample, the COD reduction is 51.62%.

It was also observed that after 30 min of the reaction (data not shown), the pH of the solution was stable indicating the establishment of equilibrium in the reaction. The stable pH between 3.0 and 4.0 emphasizes once again the requirement of acidic conditions conducive for better degradation of the pollutant. However, no earlier adjustments to the pH of the sample were made. The odor removal was also significant in case of both dairy wastewater and synthesized dairy wastewater.

For the TSS concentration, the reduction was 94.70% which is appreciable. Turbidity removal in case of dairy wastewater and synthesized sample was as high as 79.35 and 95.75%. The effect of residual turbidity and residual TSS concentration after treatment of synthesized wastewater is given in Figs. 13.1 and 13.2. Also, Fig. 13.3 illustrates the turbidity removal efficiency of the Fenton process on the dairy wastewater sample.

Table 13.1 Characteristics of dairy wastewater before and after treatment

Parameter	Dairy wastewater		Synthesized sample	
Treatment	Before	After	Before	After
pH	7.15	3.33	5.45	3.33
COD (mg/l)	236.27	113.31	994.42	481.04
TSS (mg/l)	NA	NA	529	27.99
Turbidity (NTU)	129.16	26.67	838	35.61
Odor	Irritant	Absent	Irritant	Mild

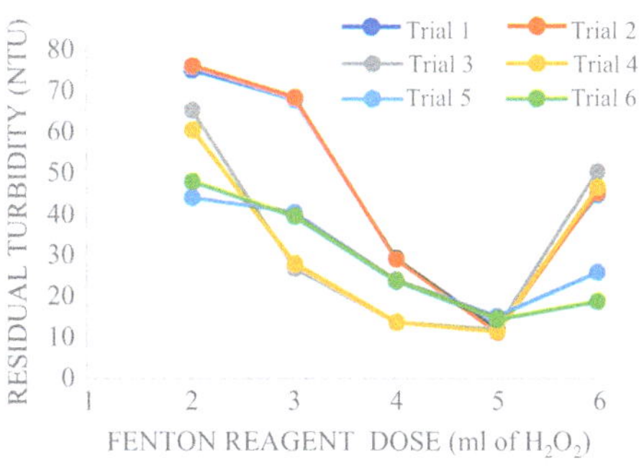

Fig. 13.1 Effect of Fenton process on turbidity removal of synthesized dairy wastewater

Fig. 13.2 Effect of Fenton process on total suspended solids removal of synthesized dairy wastewater

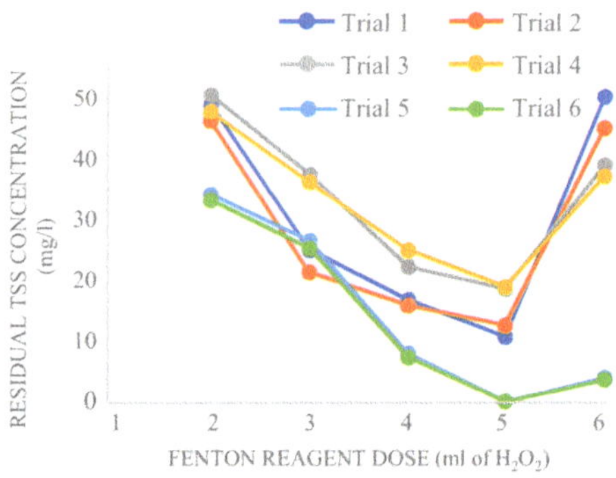

Fig. 13.3 Effect of Fenton process on turbidity removal of dairy wastewater

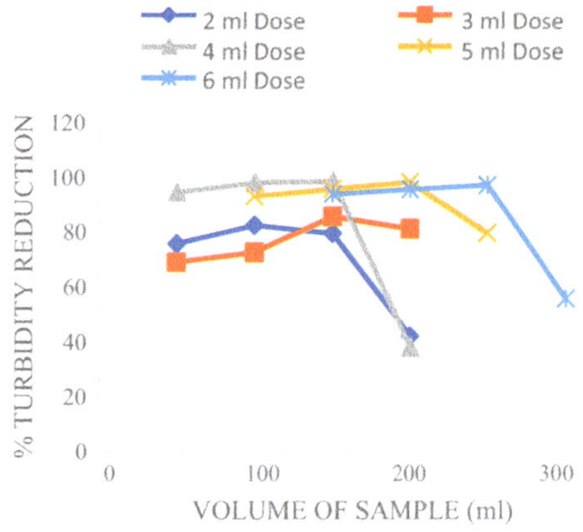

13.4 Conclusion

It should be noted that the method adopted in this study was the classical Fenton process which does not include enhancement in the reaction using UV light or any other performance improvement technologies. Having considered that, the reduction in the COD is considerable with respect to the time of treatment. However, it is advisable that the wastewater should be further treated with suitable biological methods before discharging into the receiving streams or onto the land. But, the resultant pH of the effluent is acidic which needs to be adjusted to suit the biological processes. The turbidity removal analysis of synthesized dairy wastewater shows that for the sample volume of 200 ml, dose of Fenton reagent corresponding to 5 ml achieved maximum turbidity removal. For dairy wastewater also, similar optimum dose was obtained for 200 ml sample. Total suspended solids (TSS) concentration is an important parameter of dairy wastewater which may indicate high organics. TSS removal was remarkable for the synthesized dairy wastewater samples with removal efficiency being as high as 99% for some doses.

The process of Fenton's oxidation can be effectively used by the dairy industry for giving primary oxidation treatment to the effluent before discharging into collection systems or also before further treatment systems for that matter. This is because of the fact that this oxidation process does not require specific temperature or energy. Thus, it can be concluded that the degradation of dairy wastewater using Fenton oxidation process is effective for removal of turbidity and suspended solids and can be a prospective method of treatment of dairy wastewater.

Acknowledgments The authors would like to acknowledge the assistance provided by Government Milk Processing Unit, Akola, Dist. Akola (Maharashtra) in the sample collection process. The authors would also like to express their gratitude toward Dr. S. P. Tatewar, Head, Department of Civil Engineering, Govt. College of Engineering, Amravati (Maharashtra) for his continuous support and guidance during the conduct of the study.

References

1. Saatci, Y.: Decolorization and mineralization of Remazol Red F_3B by Fenton and photo-Fenton processes. J. Environ. Eng. ASCE **136**(9), 1000–1005 (2010)
2. Gogate, P.R., Pandit, A.B.: A review of imperative technologies for wastewater treatment I: oxidation technologies at ambient conditions. Adv. Environ. Res. **8**, 501–551 (2004)
3. Loures, C.C.A., Samanamud, G.R.L., de Freitas, A.P.B.R., Oliveira, I.S., de Freitas, L.V., Almeida, C.R.O.: The use of advanced oxidation processes (AOPs) in dairy effluent treatment. Am. J. Theoretical Appl. Statistics **3**, 42–46 (2014)
4. Shete, B.S., Shinkar, N.P.: Dairy industry wastewater sources, characteristics & its effects on environment. Int. J. Curr. Eng. Technol. **3**(5), 1611–1615 (2013)

Chapter 14
Effect of Inlet Feed Temperature on the Performance of Direct-Ethanol Fuel Cell

Vineesh Ravi, Yohans Varghese and Caraline Ann Jacob

Abstract The present study focuses on analyzing the performance characteristics of the direct-ethanol fuel cell (DEFC) under different feed inlet temperature for ethanol solution in an anode side and air at the cathode side. The membrane used is Nafion 117 with catalyst loading on both sides. The experiment was conducted by changing the inlet feed temperature at anode and cathode side from 30–90 °C both in parallel and in cross arrangements and also varying the cell temperature from 40–80 °C. The performance of the DEFC was analyzed with a polarization curve and power density curve which was plotted by measuring voltage–current density values. From the results obtained, it was observed that for a temperature of 70 °C for feed temperature for both anode and cathode side, maximum power density of 0.00416 W/cm^2 was obtained and maximum power density of 0.00468 W/cm^2 was obtained when the combination of temperature was 30 and 90 °C inlet feed and air, respectively. Increase in cell temperature increased the performance and maximum was obtained at 0.00562 W/cm^2 at a temperature of 80 °C.

Keywords DEFC · Feed temperature · Humidifier temperature · Polarization curve · Power density

14.1 Introduction

In this era, where renewable energy resources effectively replace conventional fossil fuels in applications like distributed power generation, vehicle propulsion, and rural energy services. Fuel cells, renewable energy source, generally are considered to be the best prospects in applications like small distributed power sources in remote locations such as spacecraft, weather stations, and communication centers. It showcases several other attractive characteristics like high-conversion efficiency, low-environmental pollution with extreme reliability and durability [1]. A fuel cell is an electrochemical device which converts the stored chemical energy into electrical

V. Ravi (✉) · Y. Varghese · C. A. Jacob
Department of Chemical Engineering, National Institute of Calicut, 673601 Calicut, Kerala, India
e-mail: vineeshravi@nitc.ac.in

© Springer Nature Singapore Pte Ltd. 2020

V. Sivasubramanian and S. Subramanian (eds.), *Global Challenges in Energy and Environment*, Lecture Notes on Multidisciplinary Industrial Engineering, https://doi.org/10.1007/978-981-13-9213-9_14

energy [2]. Basically, a fuel cell consists of two electrodes, i.e., anode and cathode with an electrolyte between them. Fuel is supplied to the anode, where it splits into ions and electrons. The ions move toward the cathode through the electrolyte whereas the electrons move through the external electric circuit to reach the cathode as the membrane is electrically insulated. Fuel cells are mainly of different types such as polymer electrolyte membrane fuel cell (PEMFC), molten carbonate fuel cell, solid oxide fuel cell, etc. Direct-ethanol fuel cell is a type of PEMFC, where proton-conducting membrane is used as the polymer electrolyte. In DEFC, ethanol is used as the fuel since it has a far better energy density (6.36 kWh/L) [3] and has higher efficiency and lower toxicity.

The electrocatalytic oxidation of ethanol was found that Pt-Ru or Pt-Sn was the most effective and the least poisoned [4]. The electrodes used are Pt-Ru on the anode side and Pt on the cathode side. The electrolyte used is the Nafion 117 membrane. The reaction mechanisms in the fuel cell are;

The anode reaction is

$$C_2H_5OH + 3H_2O \rightarrow 12H^+ + 12e^- + 2CO_2 \tag{14.1}$$

The cathode reaction is

$$3O_2 + 12H^+ + 12\,e \rightarrow 6H_2O, \tag{14.2}$$

The overall reaction is given by

$$C_2H_5OH + 3O_2 \rightarrow 3H_2O + 2CO_2, \tag{14.3}$$

The performance of fuel cell analyzed here is influenced by different operating parameters such as cell temperature, feed temperature, and humidification temperature of the feed. The performance of a vapor fed DEFC operating at high temperatures using a PBI-based electrolyte was studied by Lobato et al. [5], which shows temperature above 150 °C gives better performance. Pramanik and Basu [6] have studied experimentally the effect of various parameters like fuel concentration, catalyst loading, and cell temperature on fuel cell performance. Maher et al. [7] followed a parametric study of a PEMFC using three-dimensional computational fluid dynamics model. Rihabjaralla [8] investigated the parameter effects of cathode operating conditions on the performance of PEMFC using the mathematical model. The experimental investigations presented in this paper focus on the performance of DEFC under parameters such as ethanol solution feed temperature, inlet air humidification temperature, and cell temperature. Experiments have been carried out on a single fuel cell to provide systematic experimental data that may be valuable for the fuel cell developers.

14.2 Experimental Setup

The system consisted of a fuel cell assembly structure, a compressor, a humidifier unit, a peristaltic pump, a vaporizer unit, and an electronic load subject to adjustment. The ethanol solution was given as feed to the single fuel cell unit. The solution was pumped to the fuel cell by a peristaltic pump through the anode side.

Air passes through the humidifier and is supplied to the cathode side of the fuel cell. The operating conditions for the fuel cell were adjusted on the above equipment. The temperature was adjusted to the electronic load. The flow rate of ethanol solution is varied by a peristaltic pump, and the flow rate of air was regulated on the rotameter system present as part of the humidifier. These connections are demonstrated in Fig. 14.1a. The data was measured using an electrochemical work station, CHI760. The tests were performed on a single fuel cell unit. The cell consisted of a membrane electrode assembly which consists of Nafion 117 membrane with catalyst loading on both sides. The fuel cell active area was 25 cm^2. The diffusion layers employed were made of carbon paper. A catalyst loading of 4 mg/cm^2 was used on the anode side (Pt-Ru), and the same load of platinum (Pt) was used on the cathode side. The flow field is made up of graphite blocks having dimensions of 1 mm × 1mm etched with double serpentine flow channels to supplyethanol solution and air to the respective compartments. The flow channels have been illustrated in Fig. 14.1c. The MEA used in this work is depicted in Fig. 14.1d. The MEA and graphite blocks were held between aluminum and gold-plated collector plates. The margins and surface of the end plates were connected and held in place using an intricate nut and bolt system. The assembled fuel cell is shown in Fig. 14.1b.

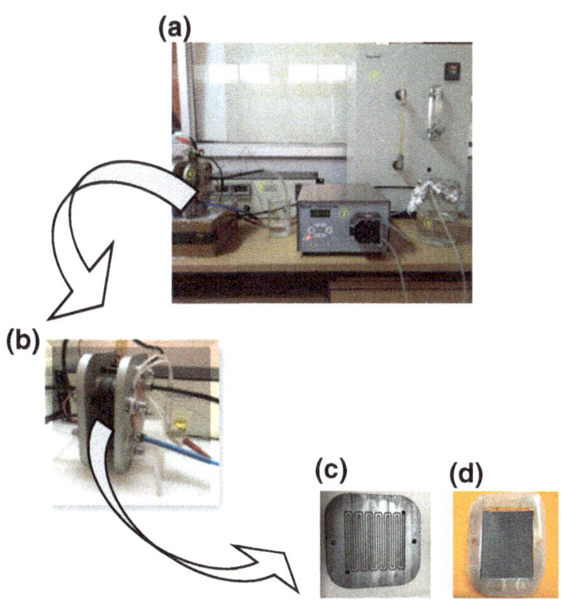

Fig. 14.1 **a** Experimental setup **b** Direct ethanol fuel cell **c** Serpentine flow field **d** Nafion membrane electrode assembly

14.3 Results and Discussion

14.3.1 Effect of Inlet Feed Temperature of Anode and Cathode Side in the Parallel Arrangement

The experiment was carried for an ethanol concentration of 2 M concentration and fuel flow rate of 1.5 ml/min with the airflow rate of 1500 ml/min. The experiment was done for 40 to 90 °C for both ethanol solution and airflow temperature. When inlet feed temperature of both temperatures was maintained at the same temperature, the polarization curve for the current density is shown in Fig. 14.2a and the plot of power density is shown in Fig. 14.2b. As observed from Fig. 14.2a, open-circuit voltage (OCV) increases from 40 to 70 °C then drops to higher temperature. The highest

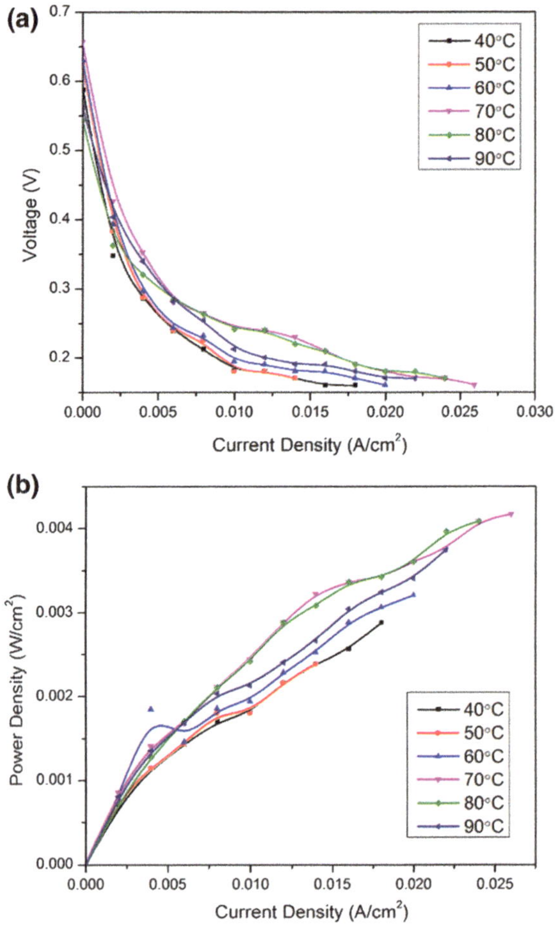

Fig. 14.2 Effect of anode and cathode fuel feed temperature on **a** voltage **b** power density of DEFC in a parallel arrangement

OCV obtained is 0.66 V at 70 °C. The steady increase in OCV value in the given interval could be attributed to the fact that as the temperature of the feed increases so does the ethanol oxidation rate increases due to increased reaction kinetics facilitated by higher feed temperature [9].

At higher temperature, the OCV value decreases because in the initial stages, when lower current is produced, lower water molecules are produced which hydrates the membrane facilitating the improved proton conductivity in the proton-conducting membrane. As the reaction progresses, more water molecules are generated which hydrates the membrane giving an increased current density at the end of the curve [10]. Another point to be noted in the graph is that the polarization curve at lower temperatures shows high OCV and at the end of the analysis, those curves show low-current densities and those curves having low-OCV values tend to provide a high-current density. This can be explained by the generation of more water molecules as the reaction progresses thus hydrating the membrane and improves the mass transfer and reaction kinetics. Likewise, at a higher temperature, the current densities decrease because at elevated temperatures dehydration of cell takes place which acts obstructively for the ethanol diffusion. Figure 14.2b shows the power density curve for parallel arrangement. From Fig. 14.2b, it was observed that the maximum power density obtained was 0.00416 W/cm^2 at a current density of 0.026 A/cm^2.

14.3.2 Effect of Ethanol Solution and Humidified Air Temperature in Cross Arrangement

The experiment was conducted by maintaining the same parameter as before that is 2 M ethanol vapor feed concentration, 1.5 ml/min of fuel flow rate and 1500 ml/min of airflow rate. The effect of inlet feed temperature of anode and cathode temperature on voltage and power density in cross arrangement is shown in Figs. 14.3a and b. As inferred from the following figures, the maximum current density is obtained when the cathode feed temperature is at 30 °C and the anode inlet temperature is at 90 °C. While the temperature at cathode increases and anode temperature decreases, the current density obtained decreases giving the lowest performance for the system having 50 °C as cathode and 70 °C as anode side temperature.

If observed closely, the pattern in which polarization curve obtained is quite the opposite for the open-circuit voltage in which cathode temperature of 40 °C and anode temperature of 80 °C give the highest OCV value. The reason for this anomaly can be explained by the reason that as the temperature of the feed increases, it increases the reaction kinetics and hence, the ethanol oxidation producing more yield. The performance of fuel cell increases with cathode inlet temperature but irrelevant as fuel temperature decreased. In the low-current region, the humidification plays a major role when the operating or feed temperature is very high; it causes evaporation inside the cell causing drying inside which reduces the mass transport of ions through the membrane so in order to keep the mass transfer smoothly. At 60 °C when both

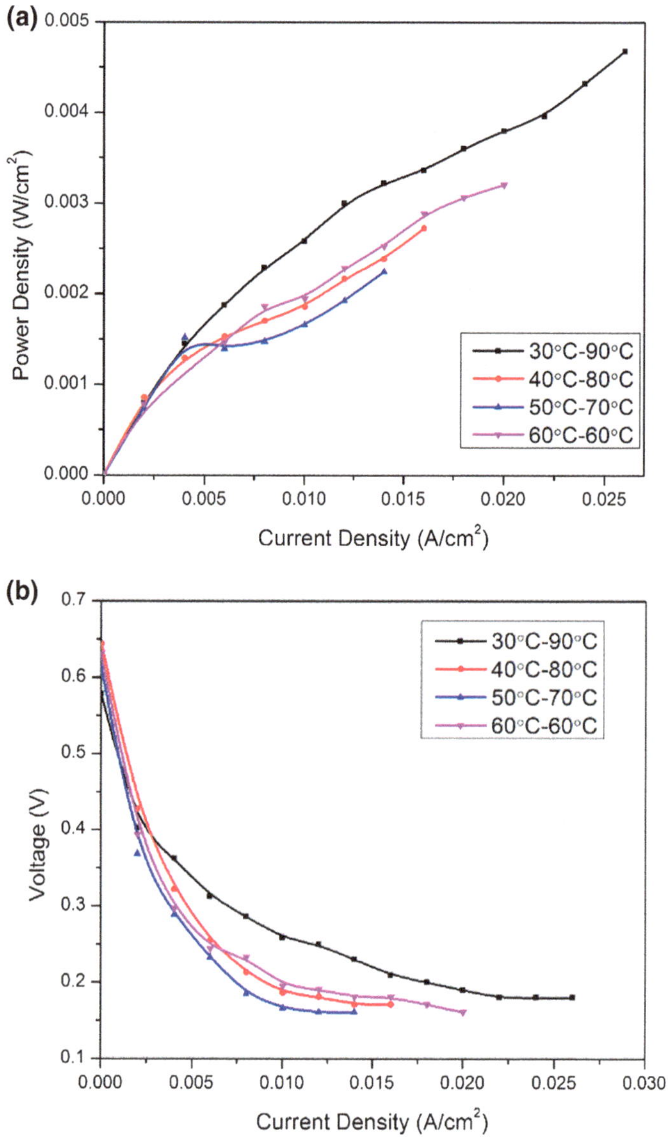

Fig. 14.3 Effect of anode and cathode fuel feed temperature on **a** voltage **b** power density of DEFC in cross arrangement

feed and cathode inlet temperature are same, it showed more uniform and pronounced effect of cathode temperature where it correlates with the previous analysis of having the maximum output at temperature 70 °C.

Same is the case, when, comparing the power density of the fuel cell; the performance is maximum when anode feed temperature is at the highest and cathode temperature plays only a minor role in the analysis. When cathode temperature increases the overall power densities, decreases as the feed temperature decreases. The maximum power density obtained here is 0.00468 W/cm^2; inlet feed temperature in the anode side is 90 °C and inlet feed temperature in the cathode side is 30 °C.

14.3.3 Effect of Cell Temperature

An increase in temperature of the fuel cell results in an increase in the performance. This is due to a reduced value of activation overpotential and the enhanced reaction kinetics of the system. The conductivity of the membrane, the electrode kinetics, and the transport properties of the cell, is all functionally dependent on temperature. Another potential change is in the diffusivity of the ethanol molecules with an increase in the temperature.

An increase in temperature could also give better OCV values. However, on increasing temperatures above 80 °C, the proton conductivity decreases. This is due to dehydration of the membrane, particularly Nafion® at elevated temperature. Figure 14.4a and b show the polarization and power density curves for various temperatures from 40–80 °C, respectively. It is obvious that a temperature of 70–80 °C seems ideal for the best performance of the DEFC.

14.4 Conclusion

As inferred from the above results, inlet feed temperature of the anode and cathode side and the cell temperature plays a crucial role in determining the maximum yield of a fuel cell. In the cross arrangement, when the anode and the cathode side inlet temperature at 90 and 30 °C, respectively, maximum power density is observed and the power density observed is 0.00468 W/cm^2. For the parallel arrangements, the maximum power density observed is at 70 °C where power is 0.00416 W/cm^2. When the parameter is based on cell temperature, the maximum power density is obtained at 80 °C which is the maximum temperature taken in the analysis. The power obtained is 0.00562 W/cm^2. It is clear from the above analysis that the power obtained when maximizing the cell temperature is far better than the power obtained by changing the inlet feed temperature of anode and cathode side. The above result is by Andreadis [11] who has observed that as cell temperature increases, so does the ethanol conversion and hence, higher power density is obtained. The cell temperature study could only be conducted until 80 °C; as at temperatures above 90 °C, the

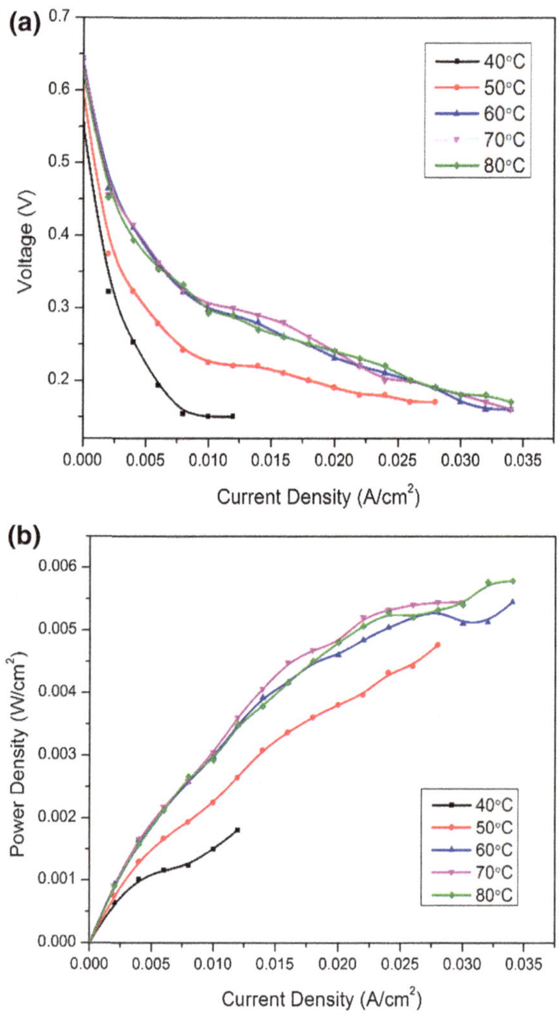

Fig. 14.4 Effect of cell temperature on **a** voltage **b** power density of DEFC

Nafion membrane starts to deteriorate. Azmy [12] in his research has pointed out that temperature variations affect all transport phenomena and reaction kinetics inside the fuel cell and also came to a conclusion that as an operating temperature increases the fuel cell, the performance improves. It can be explained that as the temperature of the feed increases, the adsorption of ethanol molecules on the catalyst increases effectively increasing the ethanol oxidation and hence, more power output. When at low current the temperature inside the cell increases with the temperature of the feed, evaporation rate increases making the cell dry and decreases the proton conductivity; humidifier temperature optimizes the temperature and the water present in the cell.

References

1. Priya, K., Sudhakar Babu, T., Balasubramanian, K., Sathish Kumar, K., Rajasekar, N.: A novel approach for fuel cell parameter estimation using simple Genetic Algorithm. Sustain. Energy Technol. Assessments **12** 46–52 (2015)
2. Yang, S., Wang, N.: A novel P systems based optimization algorithm for parameter estimation of proton exchange membrane fuel cell model. Int. J. Hydrogen Energy **37**(10), 8465–8476 (2012)
3. Badwal, S.P.S., Giddey, S., Kulkarni, A., Goel, J., Basu, S.: Direct ethanol fuel cells for transport and stationary applications—a comprehensive review. Appl. Energy **145**(October), 80–103 (2015)
4. Waszczuk, P., et al.: Adsorption of CO poison on fuel cell nanoparticle electrodes from methanol solutions: a radioactive labeling study. J. Electroanal. Chem. **511**(1–2), 55–64 (2001)
5. Lobato, J., Cañizares, P., Rodrigo, M.A., Linares, J.J.: Testing a vapour-fed PBI-based direct ethanol fuel cell. Fuel Cells **9**(5), 597–604 (2009)
6. Pramanik, H., Basu, S.: A study on process parameters of direct ethanol fuel cell. **85** 781–785 (2007)
7. Sadiq Al-Baghdadi, M.A.R., Shahad Al-Janabi, H.A.K.: Parametric and optimization study of a PEM fuel cell performance using three-dimensional computational fluid dynamics model. Renew. Energy, **32**(7) 1077–1101 (2007)
8. Jaralla, R.: Proton exchange membrane (Pem) fuel cell parametric study via mathematical modeling and numerical a Uthor'S declaration for electronic submission of a dissertation (2015)
9. El, A., Aa, M., Azmy, A.M., Sa, M.: Dynamic modelling of proton exchange membrane fuel cells for electric vehicle applications. J. Pet. Environ. Biotechnol. **5**(2), 1–7 (2014)
10. Kamarudin, M.Z.F., Kamarudin, S.K., Masdar, M.S., Daud, W.R.W.: Review: direct ethanol fuel cells. Int. J. Hydrogen Energy **38**(22), 9438–9453 (2013)
11. Andreadis, G., Stergiopoulos, V., Song, S., Tsiakaras, P.: Direct ethanol fuel cells: the effect of the cell discharge current on the products distribution. Appl. Catal. B Environ. **100**(1–2), 157–164 (2010)
12. El, A.A.A., Azmy, A.M., Mahmoud, S.A.: Effect of process parameters on the dynamic behavior of polymer electrolyte membrane fuel cells for electric vehicle applications. Ain Shams Eng. J. **5**(1), 75–84 (2014)

Chapter 15
Experimental Studies Involving Flow Visualization Over Skewed Steps and Aerospikes Using Water Channel

K. Sai Sarath, K. Dhananjayakumar, N. Sivateja and N. SaiBabu

Abstract Study on flow over skewed steps constitutes a vital branch of fluid mechanics. In the present study, the size of the wake zone behind the skewed steps and aerospikes of different shapes are visualized. Physics behind the wake zone with and without aerospike attached to the blunt-shaped object is investigated for different velocities ranges from 5 to 20 cm/s. Pressure uphill, zone of influence, flow separation and reattachment points are also observed for various skewed steps. Experimental analysis is carried out by using water channel setup which is manufactured to the designed dimensions. Work is further extended to determine drag force and coefficient of drag for skewed steps. Plots drawn with the velocities and zone of influence are analyzed with respect to different flow features. From the plots, it is clearly observed that the vortex formation is less for an object attached with a hemispherical spike.

Keywords Aerospike · Flow separation · Skewed step · Water channel · Wake zone · Zone of influence

15.1 Introduction

Frequently researchers develop various techniques to understand the physics behind the recirculation zone. Flow visualization technique is one of the techniques, which is used mostly by the researchers in order to understand the flow features in both low-speed and high-speed flow [1]. In the present paper, the same technique is applied to understand the flow features over skewed steps and aerospikes which are attached to the front nose of the various geometries. Nezu and Rodi [2] conducted a series of experiments using laser Doppler anemometer and presented the data which would help a lot for the researchers to go in-depth in understanding the turbulence intensities and also it laid a foundation for the mathematical models. Rathakrishnan [3]

K. Sai Sarath (✉) · K. Dhananjayakumar · N. Sivateja · N. SaiBabu
Department of Mechanical Engineering, Koneru Lakshmaiah Education Foundation, 522502 Guntur, India
e-mail: satyasai222@gmail.com

© Springer Nature Singapore Pte Ltd. 2020

V. Sivasubramanian and S. Subramanian (eds.), *Global Challenges in Energy and Environment*, Lecture Notes on Multidisciplinary Industrial Engineering, https://doi.org/10.1007/978-981-13-9213-9_15

conducted a set of experiments related to vortex formation and stated various definitions related to vortex length. Ahmed et al. [4] stated that spike attached to the front of blunt-nosed body will reduce the drag drastically. Khaurna and Suzuki [5–8] conducted a series of experiments and also extended their studies using aerospikes attached to blunt-nosed bodies, and they clearly explained the difficulties faced due to inertia in high-speed flows. Takama and Rathakrishnan [9] reported from their studies that varying the size of geometry leads to change in wake zone size. From their experiments, Saltzman et al. [10] concluded that there should be a reduction of drag about 40% if the square-shaped geometry is replaced with the circular shape. Sharma et al. [11] identified a twin-vortex formation behind the four plates using water tunnel apparatus. Thanigalaraju et al. [12] conducted the experiments with different geometrical tabs and reported that the formation of vortices more for flat tab as compared to circular tab. Though a lot of work is done on various noncircular geometries, work on flow features over skewed steps is limited. So, present work is concentrated on the flow features over aerospikes of different geometries attached to bluff-shaped body. Flow over skewed steps with various inclinations also discussed. Comparison has been made for all the objects with respect to variation in geometrical dimensions, velocity of flow, and angle of attack.

15.2 Experimental Methods and Material

Experiments are conducted in the fluid mechanics laboratory of K L University, Vijayawada, on various shaped skewed steps and also on the aerospikes attached to front nose of blunt-shaped body. Flow over the skewed steps and aerospikes attached to front nose of blunt-shaped body is visualized using a water channel. In the present paper, water is used as working fluid since the nature of behavior of water and air is same when the flow is incompressible. A schematic and pictorial view of the experimental unit is shown in Fig. 15.1. Water will flow from the overhead tank and then flow through the stagnation chamber and spill over the wedge. The downstream of the wedge is provided with 3 stainless steel wire screens having the holes of 4, 3, and 2 mm diameters, respectively in order to maintain the uniform flow velocity. The models to be tested are placed in front of wire screens.

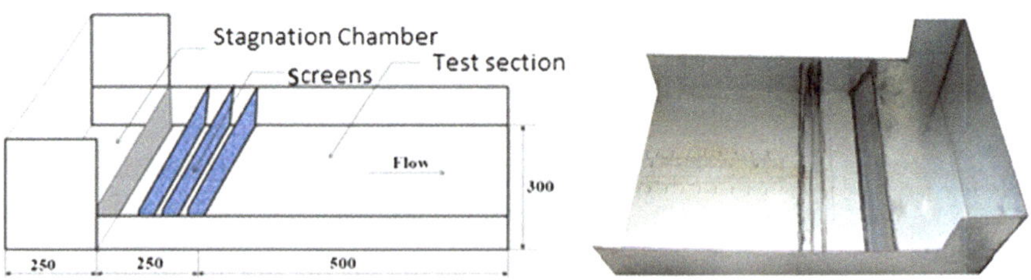

Fig. 15.1 Schematic and pictorial view of the experimentation unit

Two-skewed steps models are having the dimensions of length × height of 50 mm × 10 mm and 55 mm × 12 mm, respectively are fabricated as shown in Fig. 15.2. Experiments are conducted with different inclinations normal to the flow direction as shown in Fig. 15.3. Visualization analysis is done for each model to measure the flow velocity by floating particle method. All the experiments are performed with velocities of 5, 10, 15, and 20 cm/s. For better visibility of flow, potassium permanganate color die is mixed with water by injecting it in front of the wire screens.

Fig. 15.2 Pictorial view of tested specimens

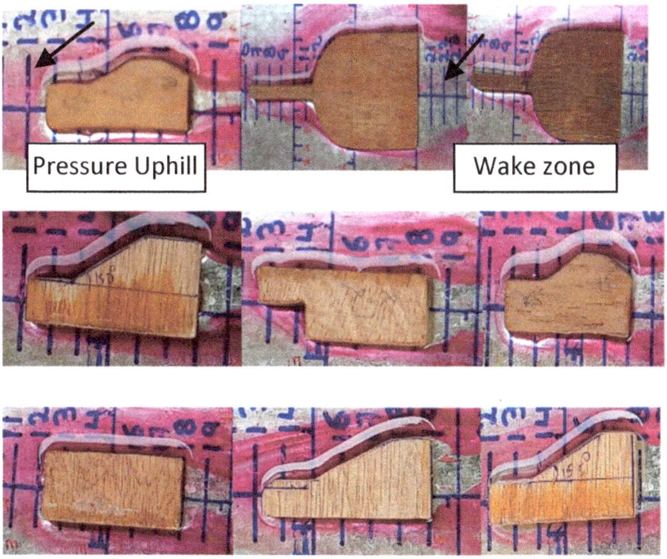

Fig. 15.3 Flow visualization over the tested objects

All flow patterns and their features (as shown in Fig. 15.3) over various skewed step model are captured with a video camera to measure various parameters such as pressure uphill, zone of influence, recirculation zone size, and vortex formation for different flow velocities ranging from 5 to 20 cm/s. Further, the work is extended for visualization of the flow features over aerospikes of different shapes attached to the front nose of the blunt body. Drag force and coefficient of drag are estimated by mathematical calculation for various models.

15.3 Results and Discussion

It is very essential to understand the flow features such as pressure uphill, zone of influence, and recirculation zone size over the skewed steps. And also it is important to estimate the drag force over aerospikes. It is also equally important to visualize the flow attachment and separation point over all the noncircular objects.

Figure 15.4 shows the variation of pressure uphill size with flow velocity. It is clearly understood that for bluff body and stepped model with 90° inclination, pressure uphill size is increasing up to the velocity is 15 cm/s. It occurs due to the creation of positive pressure ahead of the object. Then pressure uphill size decreases till velocity reaches 20 cm/s. This is because of the reduction in the wake zone size behind the object and also due to flow separation occurs behind the object. Comparing all the models, it is observed that pressure uphill size is less for the step with 20° inclination. This happens due to the formation of less number of eddies behind the object.

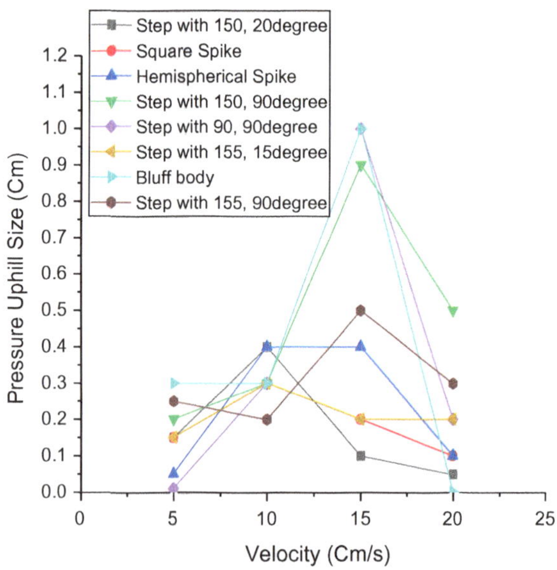

Fig. 15.4 Variation of pressure uphill with velocity

Figure 15.5 shows the variation of recirculation zone size with the flow velocities. It is observed that for all the models, the recirculation zone size is increasing till velocity approaches 15 cm/s. It is very interesting to see the flow pattern over hemispherical spike. As flow approaches the spike, it changes its direction and the recirculation zone is forming at behind the spike and size of the zone is very less for all the velocities except at 5 cm/s. This leads to a reduction in drag force. From Fig. 15.4, flow attachment and flow detachment can be visualized.

Figure 15.6 shows the variation of zone of influence with flow velocity. It is

Fig. 15.5 Variation of Recirculation zone size with velocity

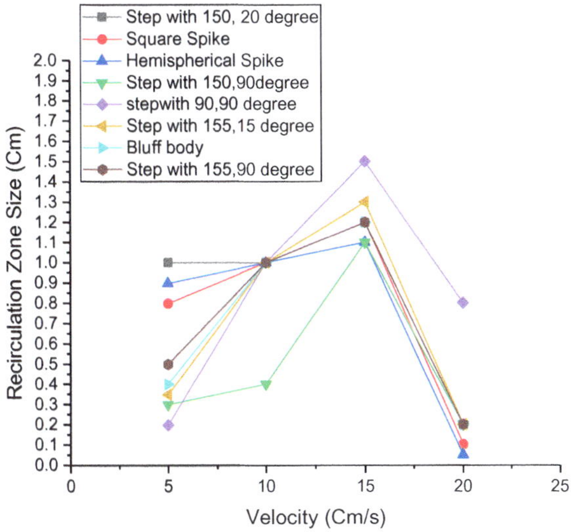

Fig. 15.6 Variation of zone of influence with velocity

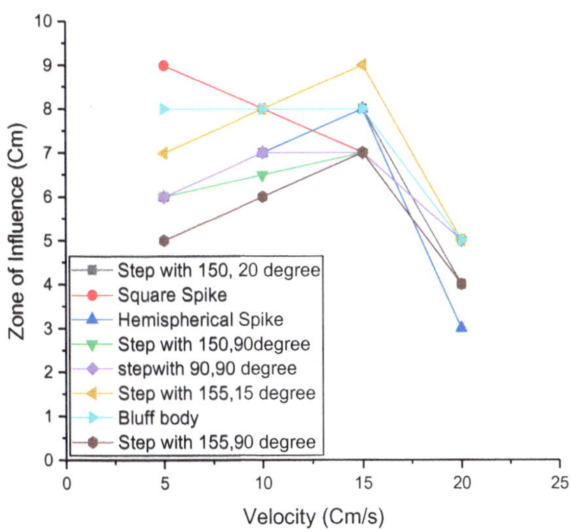

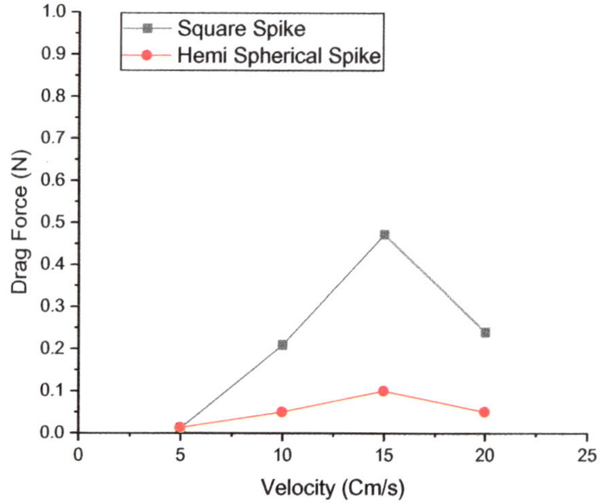

Fig. 15.7 Variation of drag force with velocity

observed that the zone of influence is less up to velocity of 5 cm/s and then increases marginally till velocity increases 15 cm/s. And then decreases up to 20 cm/s. It gives clear indication of a reduction in drag force.

Figure 15.7 shows the influence of drag force with flow velocities for aerospikes models (both square and hemispherical spike). It is observed that comparing both square and hemispherical spike, reduction in drag force is much lower with hemisphere spike. Drag force tends to increase with increase in velocity till 15 cm/s. After that, the drag force decreases to 20 cm/s. This is due to the decrease in recirculation zone size behind the hemispherical spike.

15.4 Uncertainty Analysis

Floating particle method is used to measure the velocity of flow and the error in the measurement predicted to be less than ±2%. All the features like zone of influence, pressure uphill, and recirculation zone size are about ±3% and uncertainty in the dimensions as compared to the values taken from the images.

15.5 Conclusions

Flow visualization over skewed steps and aerospikes is done for various velocities ranging from 5 to 20 cm/s. Pressure uphill size, zone of influence, recirculation zone size, and Drag force increase up to 15 cm/s. Thereafter, all the parameters are decreasing up to 20 cm/s. This shows that the velocity of flow above 15 cm/s gives better results. Hemispherical spike will possess less drag, which gives positive output regarding the change in dimensions of the objects. The present study can be extended for a different set of models with variation in dimensions and also variation of range of velocity of flow.

References

1. Kruthiventi, S., Rayapati S., Nikhil S.: Manideep. Experimental studies involving flow visualization over non-circular geometries. In: Regupathi, I., Shetty, K.V., Thanabalan, M. (eds.) Recent Advances in Chemical Engineering pp. 21–29. Springer, Singapore (2016
2. Nezu, I., Rodi, W.: Open-channel flow measurements with a laser Doppler anemometer. J. Hydraul. Eng. **112**(5), 335–355 (1986)
3. Rathakrishnan E.: Visualization of the flow field around a flow plate. Instrum. Measur. Mag. IEEE **15**(6), 8–12 (2012)
4. Ahmed M.Y.M., et al.: Recent advances in the aerothermodynamics of spiked hypersonic vehicles. Prog. Aerosp. Sci. **49**, 425449 (2011)
5. Khurana, S., Suzuki, K., Assessment of aerodynamic effectiveness for aerospike application on hypothesized lifting body in hypersonic flow. In: 31st A.I.A.A. Applied Aerodynamics Conference (2013)
6. S. Khurana, K. Suzuki, *Application of Aerospikes for LiftingBody Configuration in Hypersonic Flow at Mach 7, 43rd A.I.A.A. Fluid Dynamics Conference and Exhibit (2013)*
7. Khurana, S., Suzuki, K.: Towards heat transfer control by aerospikes for lifting-body configuration in hypersonic flow at Mach 7. In: 44th A.I.A.A. Thermophysics Conference (2013)
8. Khurana, S., et al.: Flow field behaviour with Reynolds number variance around a Blunted-nosed Body with Spike. Mod. Phys. Lett. B **30**, 1–13 (2016)
9. Takama, Y., Suzukiand, K., Rathakrishnan, E.: Visualization and size measurement of Vortex shed by flat and arc plates in an uniform flow. Int. Rev. Aerospace Eng. (IREASE) **1**, 55–60 (2008)
10. Saltman, E.J., Meyer, R.R.: Drag reduction obtained by rounding vertical corners of box shaped vehicle. NASA Flight research center March (1974)
11. Sharma, H.,Vashishtha, H.A., Rathakrishnan, E.: Twin-vortex flow physics. Proc. I MechE Part G J. Aerospace Eng. **222**, 783–788 (2008). https://doi.org/10.1243/09544100jaero322
12. Thanigalaraju, S., Jayaprakash, S., Elangovan, S., Rathakrishnan, E.: Influence of tab geometry and its orientation on underexpanded sonic jets. Proc. IMech E, Part G J. Aerospace Eng. **222**(G3), 331–339 (2008)

Chapter 16
Study on the Effective Removal of Chromium VI via Polysulfone/TiO_2 Nanocomposite Membranes and Its Antifouling Property

Lovey Jacob, Tiza Moses, Shiny Joseph, Lity Alen Varghese and Shankar Nalinakshan

Abstract By thermal induced phase inversion technique, polysulfone/TiO_2 (psf/TiO_2) nanocomposite membranes are prepared with N-methyl pyrrolidone(NMP) as solvent and polyvinylpyrrolidone(PVP) as the porogen. The prepared membranes are then undergone physical, morphological, thermal characterization studies using a universal testing machine, scanning electron microscope, thermogravimetric analyzer, and X-ray diffractometer. Membrane properties such as porosity, equilibrium moisture content, contact angle, flux rate, % rejection, and its antifouling property are also determined. Polysulfone nanocomposite membranes with 1% TiO_2 is found to have a higher porosity, percentage of water uptake, mechanical properties, and a lower contact angle. Polysulfone with 1% TiO_2 is found to have a higher performance with the flux rate of 75.714 l/m^2 h along with the 94.045% Cr(VI) ion rejection and a better antifouling property.

Keywords Ultrafiltration · Phase inversion · Nanocomposite membranes · TiO_2 nanoparticle · Antifouling property

16.1 Introduction

In recent years, amount of highly toxic contaminants such as organic compounds, dyes, heavy metals, and pesticides increases in the water resources due to human

---

L. Jacob · S. Joseph (✉) · L. A. Varghese
Department of Chemical Engineering, National Institute of Technology Calicut, 673601 Calicut, Kerala, India
e-mail: shiny@nitc.ac.in

T. Moses
Department of Civil Engineering, UKF College of Engineering and Technology, 691302 Kollam, Kerala, India

S. Nalinakshan
Department of Electrical & Electronics Engineering, The National Institute of Engineering, 570008 Mysuru, Karnataka, India

© Springer Nature Singapore Pte Ltd. 2020

V. Sivasubramanian and S. Subramanian (eds.), *Global Challenges in Energy and Environment*, Lecture Notes on Multidisciplinary Industrial Engineering, https://doi.org/10.1007/978-981-13-9213-9_16

activities. These contaminants adversely affect the human health, and the separation of these contaminants from water is found to be a complex process. Among all the industrial wastes, tannery effluents are ranked as the highest pollutants and more than 90% of leather production industries use chromium salts as tanning substances [1]. Along with other pollutants, tannery effluent contains organic foulants which may originate from the raw material or may introduce during the tanning processes [2]. Various techniques were used to remove these contaminants from water. Among these techniques, membrane separation is found to be more effective due to their low capital cost, compactable design, and low energy consumption [3]. The main disadvantage of this technology is that membranes are susceptible to fouling. Polymeric materials are still widely used materials for the preparation of membranes due to its high thermal, mechanical, and chemical stability, good flexibility. Among different polymeric materials, polysulfone is widely used for the water treatment applications due to its high film-forming nature, low-cost, high mechanical and thermal stability [4]. Since the hydrophobic nature of polysulfone results in severe fouling of the membranes, several studies on the modifications of psf membranes are carried out by the researchers to enhance the hydrophilicity of the membranes. From these studies, it was found that the incorporation of inorganic particles such as ZrO_2, TiO_2, SiO_2, nanoclay [5–8] gives stable and better performances with improved hydrophilicity and antifouling properties [9]. Today, many studies are focused on the incorporation of TiO_2 nanoparticles due to its self-cleaning, antibacterial, and photocatalytic properties [6]. Addition of polyvinyl pyrrolidone will act as a porogen which increases the porosity, thereby enhancing the membrane flux rate [10].

16.2 Experimental

16.2.1 Materials

Polysulfone pellets were purchased from Sigma-Aldrich, USA; N-methyl pyrrolidone and polyvinyl pyrrolidone were purchased from Otto chemicals, India, and TiO_2 nanoparticles from SRL, India.

16.2.2 Preparation of Polysulfone Membrane and Polyamide Membranes

An adequate amount of polysulfone with solvent N-methyl pyrrolidone was taken in a beaker and placed on a magnetic stirrer at 60 °C. The required amount of PVP and TiO_2 is taken and sonicated in a probe sonicator for the complete dispersion in the solvent. This sonicated nanoparticle is fed into the beaker after all the polysulfone pellets are dissolved in the solvent. A thin-film applicator is used to draw the solution into desired thickness of 200 μm membrane on a glass plate, and it was immediately dipped into the distilled water and kept in distilled water for 24 h for complete phase inversion process.

16.2.3 Characterization Studies

Characterization studies were conducted for both neat and nanocomposite membranes. The prepared membranes were tested for mechanical properties in a Shimadzu Autograph Universal Testing Machine (UTM) according to ASTM standards. SEM images were taken for analyzing surface morphology, and thermal characterizations were done by TGA. Hydrophilicity measurement is carried out using goniometer.

The porosity and % water intake were calculated using the equation

$$P(\%) = (W_1 - W_o)/(\rho \mathrm{Ah}) * 100 \quad (16.1)$$

$$\mathrm{EWC}(\%) = \frac{(W_1 - W_o) \times 100}{W_1} \quad (16.2)$$

where P is the membrane porosity (%), ρ is the density of water 998 kg/m^3, A is the membrane surface area (m^2), h is the thickness of the membrane (m), and W_1 is wet weight of the membrane and W_o dry weight of the membrane (kg), respectively [11].

16.2.4 Filtration Experiment

The performance and the antifouling properties of the membranes were conducted in a cross-flow filtration unit. The effective area of the system is 42 cm^2. Filtration unit can be operated under the maximum pressure of 69 bar and the temperature up to 90 °C. The performance of the fabricated membrane is evaluated using 5 ppm chromium VI in distilled water (DI). The flux rate and % rejection are calculated using the following equation

$$J_0 = \Delta V/\mathrm{Am} * \Delta t \quad (16.3)$$

$$R_0(\%) = (1 - C_p/C_f) \times 100 \quad (16.4)$$

where C_p is the concentration of permeate solution and C_f is the concentration of feed solution, and where J_0 is the water flux in l/m^2 h, Am is the effective area of membrane in m^2, ΔV is the volume of permeate solution in m^3, and Δt is the time in h [12].

Antifouling performance test was conducted using organic foulant BSA. The experiment is conducted with 100 mg/l of BSA/DI solution, and flux rate is calculated (J_1). The membrane is then taken out for flushing, and then again conducted the filtration test with distilled water DI(2) and the corresponding flux rate is noted as J_2. The antifouling performance is evaluated by calculating flux recovery ratio (FRR), irreversible resistance (R_{ir}), reversible resistance (R_r), and total fouling (R_t) [13] of membrane fouling using equations,

$$\text{FRR}\,\% = (J_2/J_0) * 100 \quad (16.5)$$

$$R_r\,\% = [(J_2 - J_1)/J_0] * 100 \quad (16.6)$$

$$R_{\text{ir}}\,\% = [(J_0 - J_2)/J_0] * 100 \quad (16.7)$$

$$R_t\,\% = [(J_0 - J_1)/J_0] * 100 \quad (16.8)$$

where J_0 is the initial flux rate.

16.3 Results and Discussions

16.3.1 Effect of Concentration of TiO_2 on the Mechanical Property of Nanocomposite Membranes

See Fig. 16.1.

Here, the weight composition of TiO_2 nanoparticles varies from 0 to 2% and the mechanical properties were evaluated. From the above graph, it was found that tensile

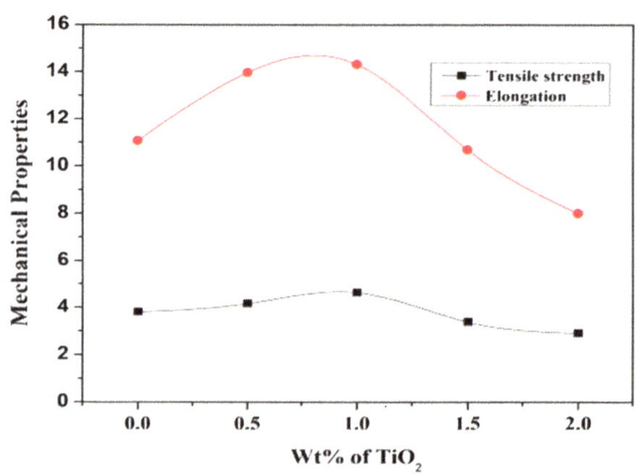

Fig. 16.1 Mechanical properties for the neat and nanocomposite membranes

strength and % elongation of the nanocomposite membrane with 1% TiO_2 are higher than the neat membrane. As the amount of TiO_2 nanoparticles increases, the mechanical strength goes on increases to 1% and then starts decreasing. This improvement in mechanical properties is due to the strong interfacial adhesion between TiO_2 nanoparticles and the polymer matrix. Further addition of TiO_2 nanoparticles decreases the membrane strength which may due to a lesser affinity between TiO_2 nanoparticles and polysulfone [14].

16.3.2 Effect on Porosity, %Water Uptake, and the Contact Angle of Nanocomposite Membranes by Varying the TiO_2 Concentration

Membrane hydrophilicity and membrane porosity have a close relationship with the membrane performance. Contact angle measurement directly relates to the hydrophilicity of the membranes. A lower contact angle measurement corresponds to higher hydrophilicity. From Figs. 16.2 and 16.3, it was found that porosity and water uptake increases with increase in the TiO_2 nanoparticle content up to 1% and then starts decreasing. Likewise, contact angle decreases with increase in TiO_2 nanoparticle content up to 1% which corresponds to higher hydrophilic nature of nanocomposite membranes and beyond 1% contact angle increases. Increase in the porosity is due to better solvent—nonsolvent diffusion which resulted in better hydrophilicity. Above 1% viscosity overrule the rate of diffusion resulted in lower water uptake and porosity and higher contact angle [4].

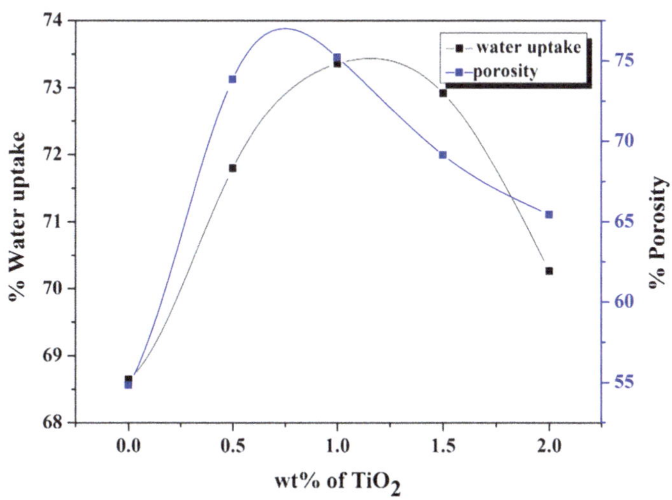

Fig. 16.2 % water uptake and porosity for nanocomposite membranes with varying TiO_2 nanoparticle content

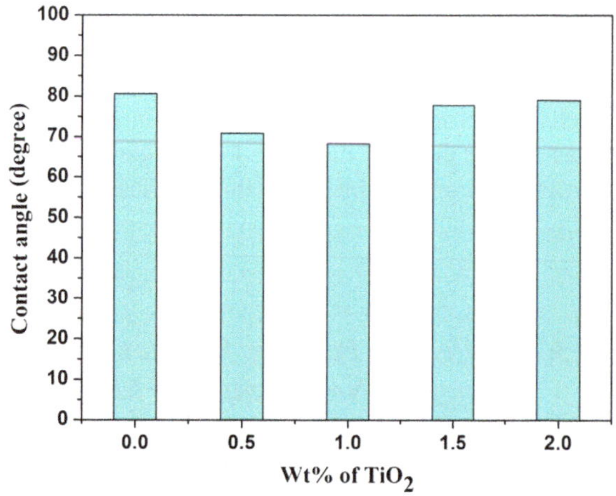

Fig. 16.3 Contact angle for nanocomposite membranes with varying TiO_2 nanoparticle content

16.3.3 Effect of TiO_2 on the Performance of the Membranes

Membrane performance study was conducted in a cross-flow filtration unit. Figure 16.4 shows flux rate and % rejection of chromium VI of nanocomposite membranes. From the figure, it was found that the nanocomposite membranes which possessed higher porosity and hydrophilicity, possessed a higher flux rate. Membrane surface entrapped with 1% TiO_2 nanoparticle is more hydrophilic due to the higher affinity of metal oxide to water. This resulted in a higher flux rate of the nanocomposite membranes [15]. Above 1% TiO_2 nanoparticle addition, it will clog

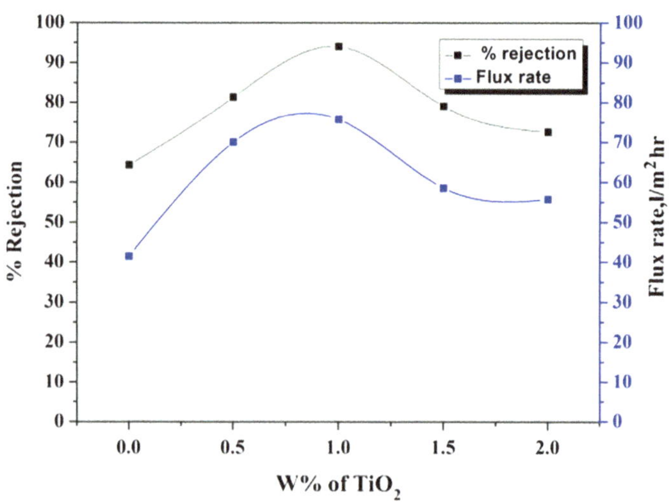

Fig. 16.4 Flux rate and rejection performance of nanocomposite membranes with varying TiO_2 nanoparticle content

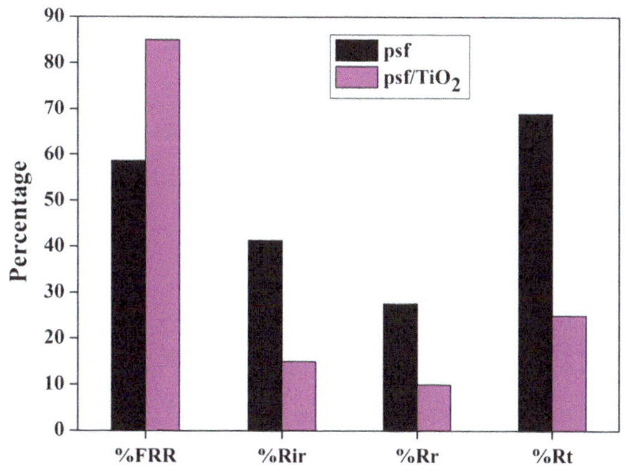

Fig. 16.5 Antifouling properties of psf and nanocomposite membrane

the membrane pores, thereby decreasing the flux rate. % rejection of Cr (VI) is found to be higher for the nanocomposite membranes with 1% TiO_2, and above that, it starts decreasing. The rejection of Cr ions is due to the adsorption of chromium ions on to the surface of the membranes. Here, the rejection of chromium (VI) for nanocomposite membrane is found to be higher than neat polysulfone membranes.

16.3.4 Antifouling Performance Study Using Organic Foulant (BSA)

See Fig. 16.5.

The antifouling performance study was conducted using organic foulant (BSA) for 1% TiO_2 nanocomposite membrane and neat membrane. As the filtration is carried out, the flux rate decreases for both nanocomposite membrane and neat membrane due to the adsorption of BSA on the membranes. Since the nanocomposite membranes are more hydrophilic than the neat membranes, the interactions between the membrane surface and BSA would not be so strong so that it can easily remove from the surface which results in the higher flux recovery ratio and a lower flux loss. Here, % FRR increases from 58 to 85% and % total resistance decreases from 68 to 39%.

16.3.5 Morphological Characterization

16.3.5.1 SEM Micrographs

See Fig. 16.6.

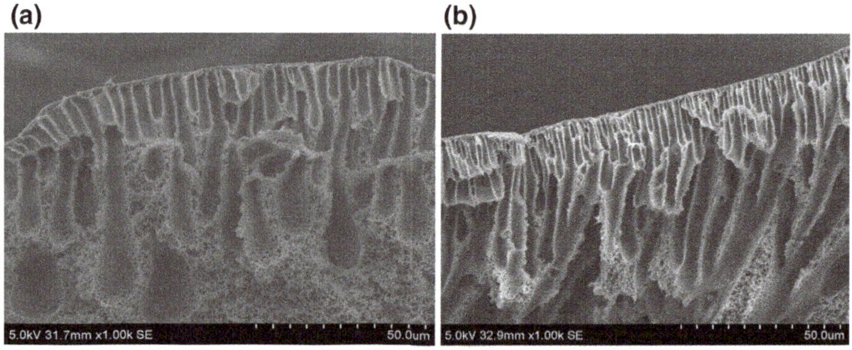

Fig. 16.6 SEM images of **a** neat membrane and **b** nanocomposite membrane

The figure shows the cross-sectional view of nanocomposite and neat membranes. From the SEM micrographs, it was found that the microvoid channel length increases by the addition of TiO_2 nanoparticles as compared to neat membranes. The improved solvent–nonsolvent interaction resulted in the enhancement of pores in the sublayer.

16.3.5.2 XRD

See Fig. 16.7.

From the XRD pattern, d-spacing value is calculated using Bragg's law and it was found that *d*-spacing value for nanocomposite membrane is higher than the neat membrane. Higher *d*-spacing value confirms exfoliated membrane formation which in turn indicates higher hydrophilicity of the membranes.

Fig. 16.7 XRD images of neat psf membrane and 1% TiO_2 nanocomposite membrane

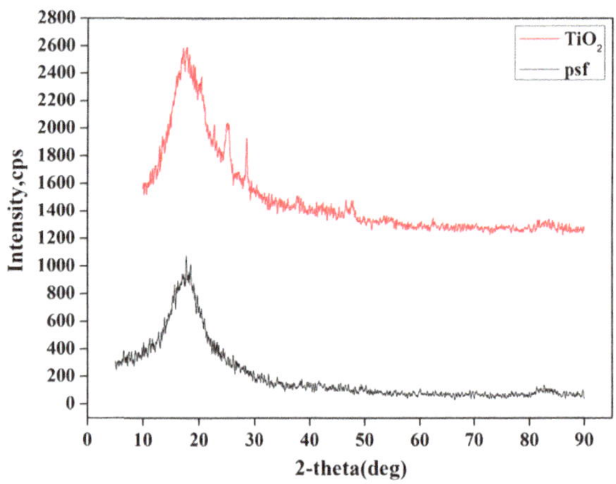

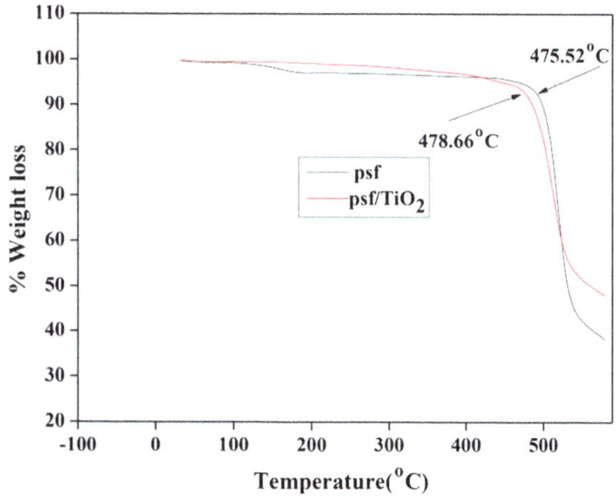

Fig. 16.8 TGA curves of neat psf membrane and 1% TiO_2 nanocomposite membrane

16.3.5.3 Thermogravimetric Analysis

See Fig. 16.8.

TGA curves of the psf membrane and nanocomposite membrane show that the peak degradation temperature of pure polysulfone membrane is 475.52 °C and for the nanocomposite membranes is 478.66 °C. This shows that the thermal stability of the nanocomposite membrane is found to be higher than that of pure polysulfone membrane.

16.4 Conclusion

In this work, ultrafiltration polysulfone membranes with the incorporation of varying TiO_2 nanoparticle content were prepared and performance studies were conducted for efficient chromium removal. Here, morphological characterisation studies show that the incorporation of TiO_2 nanoparticle increases the length of the microvoid channels which indicates the hydrophilicity of the membrane. XRD pattern reveals the information about the formation of an exfoliated membrane which also gives the indication of hydrophilicity of the nanocomposite membranes. TGA curves show that the addition of TiO_2 nanoparticle improves the thermal stability of the membrane. From the filtration studies, it was found that nanocomposite membranes with 1% TiO_2 nanoparticles are having a higher flux rate of 75 l/m^2 h and a higher % rejection of 94% along with higher antifouling performance when compared to neat polysulfone membranes.

References

1. Belay, A.A.: Impacts of chromium from tannery effluent and evaluation of alternative treatment options. J. Environ. Prot. **1**(01), 53 (2010). https://doi.org/10.4236/jep.2010.11007
2. Islam, B.I., Musa, A.E., Ibrahim, E.H., Salma, A.A.S., Babiker, M.E.: Evaluation and characterization of tannery wastewater. J. Forest Prod. Indust. **3**(3), 141-150 (2014). ISSN: 2325–4513 (print) ISSN 2325–453x (online)
3. Lin, J., Ye, W., Zhong, K., Shen, J., Jullok, N., Sotto, A., Van der Bruggen, B.: Enhancement of polyethersulfone (PES) membrane doped by monodisperse Stöber silica for water treatment. Chem. Eng. Process. **107**, 194–205 (2016)
4. Yang, Y., Zhang, H., Wang, P., Zheng, Q., Li, J.: The influence of nano-sized TiO_2 fillers on the morphologies and properties of PSF UF membrane. J. Membr. Sci. **288**, 231–238 (2007)
5. Gene, I., Kuypers, S., Leysen, R.: Effect of the addition of ZrO_2 to polysulfone-based oF membrane. J. Membr. Sci. **11**, 334–337 (1996)
6. Damodar, R.A., You, S.J., Chou, H.H.: Study the self cleaning, antibacterial and photocatalytic properties of TiO_2 entrapped PVDF membranes. J. Hazard. Mater. **172**(2–3), 1321–1328 (2009)
7. Bottino, A., Capannelli, G., D'Asti, V., Piaggio, P.: Preparation and properties of novel organic–inorganic porous membranes. Sep. Purif. Technol. **22–23**, 269–275 (2001)
8. Mahmoudian, Mehdi: Peyman Gozali Balkanloo, Clay-hyperbranched epoxy/polyphenylsulfone nanocomposite Membranes. Iran. Polym. J. **26**, 711–720 (2017)
9. Ong, C.S., Lau, W.J., Goh, P.S., Ng, B.C., Ismail, A.F.: Preparation and characterization of PVDF–PVP–TiO_2 composite hollow fibre membranes for oily wastewater treatment using submerged membrane system. Desalination Water Treatment **53**, 1213–1223 (2015)
10. Yoo, S.H., Kim, J.H., Jho, J.Y., Won, J., Kang, Y.S.: Influence of the addition of PVP on the morphology of asymmetric polyimide phase inversion membranes: effect of PVP molecular weight. J. Membr. Sci. **236**, 203–207 (2004)
11. Khalid, A., Al-Juhani, A.A., Al-Hamouz, O.C., Laoui, T., Khan, Z., Atieh, M.A.: Preparation and properties of nanocomposite polysulfone/multi-walled carbon nanotubes membranes for desalination. Desalination **367**, 134–144 (2015)
12. Emadzadeh, D., Lau, W.J., Rahbari-Sisakht, M., Daneshfar, A., Ghanbari, M., Mayahi, A., Matsuura, T., Ismail, A.F.: A novel thin film nanocomposite reverse osmosis membrane with superior anti-organic fouling affinity for water desalination. Desalination **368**, 106–113 (2015)
13. Wang, Y., Wang, T., Su, Y., Peng, F., Wu, H., Jiang, Z.: Remarkable reduction of irreversible fouling and improvement of the permeation properties of poly (ether sulfone) ultrafiltration membranes by blending with Pluronic F127. Langmuir **21**, 11856–11862 (2005)
14. Cui, A., Liu, Z., Xiao, C., Zhang, Y.: Effect of micro-sized SiO_2-particle on the performance of PVDF blend membranes via TIPS. J. Membr. Sci. **360**, 259–264 (2010)
15. Bae, T.H., Tak, T.M.: Effect of TiO_2 nanoparticles on fouling mitigation of ultrafiltration membranes for activated sludge filtration. J. Membr. Sci. **249**, 1–8 (2005)

Chapter 17
Modelling and Optimization of Industrial Scale Membrane Steam Reformer for Production of Hydrogen

S. Arun Senthil, N. Arun Prem Anand and S. Sundaramoorthy

Abstract With increasing demand for hydrogen as an environmentally friendly fuel, the catalytic steam reforming of natural gas is gaining prominence as the most attractive processes for production of hydrogen. In the conventional catalytic steam reformer (CSR), the maximum achievable conversion is limited by equilibrium. This limitation is overcome in a membrane steam reformer (MSR), in which the equilibrium is shifted towards higher methane conversion by continuously removing hydrogen from the reaction site and letting it permeate out through a palladium membrane. In spite of this advantage, no industrial scale application of MSR has so far been reported and this is mainly due to high cost of preparation of Pd membrane. In this work, a mathematical model is developed to represent the industrial scale operation of a MSR. The model is simulated to analyse the effect of various operating parameters on performance of MSR. An optimization study is carried out to obtain the optimum values of operating parameters at which the methane conversion is maximum. Effect of increasing the area of Pd membrane on the performance of MSR is studied. It is shown in this work that it is possible to achieve up to 99.9% methane conversion in MSR.

Keywords Catalytic steam reformer · Clean energy · Hydrogen production · Membrane steam reformer · Modelling · Optimization

17.1 Introduction

The demand for hydrogen as a clean fuel will be on the rise in future, when fuel cell technology becomes an economically viable alternative for large-scale generation of

S. A. Senthil · N. A. P. Anand
Department of Chemical Engineering, Sri Venkateswara College of Engineering, 602117 Sriperumbudur, India

S. Sundaramoorthy (✉)
Department of Chemical Engineering, Pondicherry Engineering College, 605014 Pondicherry, India
e-mail: ssm_pec@yahoo.com; ssm_pec@pec.edu

© Springer Nature Singapore Pte Ltd. 2020

V. Sivasubramanian and S. Subramanian (eds.), *Global Challenges in Energy and Environment*, Lecture Notes on Multidisciplinary Industrial Engineering, https://doi.org/10.1007/978-981-13-9213-9_17

electric power. With abundant availability of natural gas resources across the globe, catalytic steam reforming (CSR) process is expected to gain prominence as an attractive process for conversion of methane to hydrogen. However, maximum achievable conversion in a conventional catalytic steam reforming process is limited to the equilibrium conversion, which is reported to be around 60% only in industrial operations [1]. Thus, there is a need to intensify the steam reforming process and achieve a higher methane conversion for meeting higher hydrogen demand projected in future. One of the ways by which this can achieved is to place a palladium membrane in the catalytic reactor bed, let the hydrogen permeate through it and pass out of the reaction site. This allows the reaction equilibrium to shift towards higher production of hydrogen. Thus, a conventional catalytic steam reformer (CSR) can be suitably modified into a membrane steam reformer (MSR) by placing a Pd membrane in it.

A number of experimental and simulation studies have been reported in the literature [2] on laboratory scale membrane steam reformers. However, there is no evidence in the literature on any industrial scale application of MSR technology. As of now, high cost of preparation of Pd membrane [3] is a reason why MSR technology has not found its place in industry for production of hydrogen. However, cost of palladium membrane may be expected to come down in future as the number of industrial applications of MSR goes up. In the absence any industrial scale application of MSR reported at present, this work addresses the need to develop a mathematical model that represents the industrial scale operation of a membrane steam reformer (MSR). Such a model is essential for evaluating the feasibility of replacing the existing CSR technology by MSR technology for production of hydrogen.

In this work, a mathematical model for an industrial scale side fired catalytic steam reformer (CSR) reported in literature by Rajesh et al. [1] is taken as a reference to suitably modify and represent an industrial scale membrane steam reformer (MSR). As compared to other models reported in literature [2] that represent laboratory scale MSRs, the mathematical model developed in this work is more appropriate for analysis of industrial scale operations. Model is simulated using MATLAB to evaluate the effect of various operating conditions on methane conversion. Optimization studies are carried out to obtain the optimal values of operating parameters at which the methane conversion is maximum subject to various operational constraints and bounds.

In Sect. 17.2, details about the construction of industrial scale Membrane Steam Reformer (MSR) and the kinetic rate equations for the steam reforming reactions are presented. Detailed derivations of model equations for the MSR are presented in Sect. 17.3. Studies conducted on simulation and optimization of MSR operations are explained in Sect. 17.4. Results are presented and discussed in Sect. 17.5. Section 17.6 summarizes the conclusions drawn from this study.

17.2 Membrane Steam Reformer (MSR)

In this work, the industrial scale side fired catalytic steam reformer reported in Rajesh et al. [1] is suitably altered to represent the membrane steam reformer (MSR). The

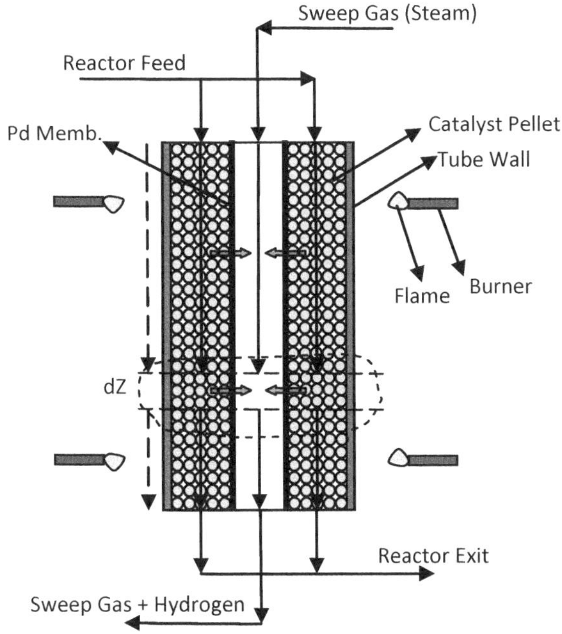

Fig. 17.1 Annular reactor tube in MSR

reformer has $n_t = 176$ numbers of annular tubes (Fig. 17.1) of length $L = 11.95$ m placed in a large refractory-lined furnace. The inner and outer diameters of the tube are, respectively, $d_i = 79.5$ mm and $d_o = 102$ mm. The annular section of the tube is the reactor filled with Ni on Al_2O_3 catalyst pellets of equivalent diameter $D_P = 17.4$ mm. The inner tube of radius r_o is an empty tube made of palladium membrane of thickness $\delta = 7.5$ μm supported on porous stainless steel. The furnace has $n_b = 112$ burners supplied with fuel and air at temperature T_g to heat the outer wall of the annular reactor tube. The feed to the reactor is passed into the annular section at a rate of F kmol/h. The feed gas is a mixture of CH_4, H_2O, recycled CO_2, recycled H_2 and N_2. F_{CH4} is the molar feed rate of methane. S/C, H/C, D/C and N/C are, respectively, the molar ratios H_2O/CH_4, H_2/CH_4, CO_2/CH_4 and N_2/CH_4 in the feed gas. H_2 produced in the annular reactor section permeates through Pd membrane wall into the inner tube at a molar flux of J_{H_2}(kmol/h/m^2) described by Sieverts law as

$$J_{H_2} = \frac{Q_o}{\delta} \exp\left(-\frac{E_P}{RT}\right)\left(\left(P y_{H_2}\right)^{0.5} - \left(P_p y_{H_2p}\right)^{0.5}\right) \tag{17.1}$$

where P is the annular section gas pressure, P_p is the permeate gas pressure, y_{H_2} is mole fraction of hydrogen in annular section and y_{H_2p} is mole fraction of hydrogen in permeate. Values of Q_o and E_p are taken from Johannessen et al. [4].

Following chemical reactions occur in the annular reactor section

$$CH_4 + H_2O = CO + 3H_2$$

$$CO + H_2O = CO_2 + H_2$$

$$CH_4 + 2H_2O = CO_2 + 4H_2$$

The kinetic rate equations r_i (kmol/kg catalyst/h) for the three reactions ($i = 1, 2, 3$) reported by Xu and Froment [5] are used in this work. The rates of formation of chemical species R_i ($i = CH_4, H_2O, CO, CO_2, H_2$) are related to r_i through reaction stoichiometric coefficients as follows

$$R_{CH_4} = -r_1 - r_3 \tag{17.2a}$$

$$R_{H_2O} = -r_1 - r_2 - 2r_3 \tag{17.2b}$$

$$R_{CO} = r_1 - r_2 \tag{17.2c}$$

$$R_{CO_2} = r_2 + r_3 \tag{17.2d}$$

$$R_{H_2} = 3r_1 + r_2 + 4r_3 \tag{17.2e}$$

17.3 Mathematical Model of MSR

A mathematical model for the industrial scale membrane steam reformer (MSR) is developed in this work by suitably modifying the model equations for a side fired industrial scale catalytic steam reformer reported by Rajesh et al. [1]. Variations of gas phase temperature, pressure and compositions in the radial direction are ignored. The solid catalyst phase is assumed to be homogeneous. The catalyst effectiveness factor η is taken as equal for all the three reactions and a value of 0.03 reported in E. Johannessen et al. [4] is assigned to it.

Model equations for annular reactor channel:

Taking over all mass balance across section dZ, we get

$$\frac{dG}{dz} = -\left(\frac{4\pi r_o L}{A_c}\right) J_{H_2} \tag{17.3}$$

Here G is mass velocity (kg/hr/m^2) of gas, A_c is cross-sectional area (m^2) of the annular reactor channel and $z = \frac{Z}{L}$ is dimensionless distance from inlet.

Taking component molar balance across section dZ, we have

$$\frac{d(v_1 C_i)}{\mathrm{d}z} = L\rho_b \eta_i R_i i = \mathrm{CH_4, H_2O, CO, CO_2} \tag{17.4}$$

$$\frac{d(v_1 C_i)}{dz} = L\rho_b \eta_i R_i - \left(\frac{2\pi r_o L}{A_c}\right) J_{\mathrm{H_2}} \quad i = \mathrm{H_2} \tag{17.5}$$

where v_1 is gas velocity (m/hr), C_i is concentration (kmol/m^3) of ith compound in the gas mixture and ρ_b is catalyst bed density (kg catalyst/m^3). Define x_i as

$$x_i = \frac{y_i}{\overline{M}} \quad i = \mathrm{CH_4, H_2O, CO, CO_2, H_2} \tag{17.6}$$

where y_i is mole fraction of ith component, $\overline{M}$ is average molecular weight written in terms of individual component molecular weights and mole fractions as,

$$\begin{aligned}\overline{M} &= y_{\mathrm{N_2}} M_{\mathrm{N_2}} + y_{\mathrm{CH_4}} M_{\mathrm{CH_4}} + y_{\mathrm{H_2O}} M_{\mathrm{H_2O}} \\ &\cdots + y_{\mathrm{CO}} M_{\mathrm{CO}} + y_{\mathrm{CO_2}} M_{\mathrm{CO_2}} + y_{\mathrm{H_2}} M_{\mathrm{H_2}}\end{aligned} \tag{17.7}$$

Writing Eqs. (17.4) and (17.5) in terms of x_i, we have

$$\frac{\mathrm{d}x_i}{\mathrm{d}z} = \frac{L\rho_b \eta_i R_i}{G} + \frac{x_i}{G}\left(\frac{4\pi r_o L}{A_c}\right) J_{\mathrm{H_2}} \quad i = \mathrm{CH_4, H_2O, CO, CO_2} \tag{17.8}$$

$$\frac{\mathrm{d}x_i}{\mathrm{d}z} = \frac{L\rho_b \eta_i R_i}{G} + \frac{(2x_i - 1)}{G}\left(\frac{2\pi r_o L}{A_c}\right) J_{\mathrm{H_2}} \quad i = \mathrm{H_2} \tag{17.9}$$

Values of y_i for any x_i is calculated by solving Eqs. (17.6) and (17.7) simultaneously.

Taking heat balance across the section dZ, we get

$$\frac{\mathrm{d}T}{\mathrm{d}z} = \frac{L}{GC_{pg}}\left[\left(\frac{\pi D_i U}{A_c}\right)(T_{wi} - T) + \rho_b \sum_{j=I}^{III} \eta_j r_j (-\Delta H_j) \cdots - \left(\frac{U_m (2\pi r_o)}{A_c}\right)(T - T_p)\right] \tag{17.10}$$

Here, T is gas temperature in outer annular reactor channel, T_P is gas temperature in inner permeate tube, D_i is inner diameter of the outer tube, C_{pg} is average specific heat capacity (kj/kg/K) of gas, $(-\Delta H_j)$ is heat of reactions (kJ/kmol) and U is annular side heat transfer coefficient (kj/h/m^2/K) calculated using the correlation reported in [1], U_m is overall heat transfer coefficient for transfer of heat occurring from the annulus to the inner membrane tube and T_{wi} is inner wall temperature of the outer annular tube calculated by solving simultaneous steady-state equations for radiative, conductive and convective heat transfer across the outer tube wall [1].

Pressure drop in the annular reactor section is calculated using Blake–Plumer equation applicable for turbulent flow of fluid trough packed bed

$$\frac{dP}{dz} = -\frac{1.75LG^2(1-\varepsilon_b)}{\varphi_s D_p \varepsilon_b^3 \rho_g} \tag{17.11}$$

Here, ϕ_s is sphericity of catalyst pellet, ε_b is bed porosity and ρ_g is density (kg/m^3) of gas.

Model equations for inner membrane tube:

Taking hydrogen mole balance across the section dZ in the inner membrane tube, we get

$$\frac{\mathrm{d}y_{H_2p}}{\mathrm{dz}} = \left(\frac{2\pi r_o L(1-y_{H_2p})}{G_p}\right) J_{H_2} \tag{17.12}$$

where G_P is molar flow rate (kmol/hr/m^2) of permeate gas (*mixture of hydrogen and steam*) per unit cross-sectional area A_P of the membrane tube

$$G_p = \left(\frac{m_s}{18}\right)\frac{1}{1-y_{H_2p}} \tag{17.13}$$

Here, m_s is mass flow rate (kg/h/m^2) of steam per unit cross-sectional area A_P of the membrane tube.

Taking steady-state heat balance across section dZ in the permeate tube, we get

$$\frac{\mathrm{d}T_p}{\mathrm{dz}} = \left(\frac{(2\pi r_o L)U_m}{G_p C_{pp}}\right)(T - T_p) \tag{17.14}$$

Here, C_{pp} is average molar specific heat capacity (kj/kmol/K) of permeate gas. Overall heat transfer coefficient U_m is related to outer annular side heat transfer coefficient U and inner tube side heat transfer coefficient h_p as

$$\frac{1}{U_m} = \frac{1}{U} + \frac{1}{h_p} \tag{17.15}$$

Dittus-Boelter equation is used for calculating inner tube side heat transfer coefficient h_p

$$h_p = 0.023\frac{k_p}{d_p}\left(\frac{\overline{G}_p d_p}{\mu_p}\right)^{0.8}\left(\frac{\overline{C}_{pp}\mu_p}{k_p}\right)^{0.33} \tag{17.16}$$

Pressure drop in the inner permeate tube is

$$\frac{\mathrm{d}P_p}{\mathrm{dz}} = -\frac{4fL}{d_p}\left(\frac{v_p^2}{2g}\right) \tag{17.17}$$

where $d_p = 2r_o$ is the membrane tube diameter and

$$v_p = \left(\frac{G_p}{A_p C_{tp}}\right) \tag{17.18}$$

$$C_{tp} = \left(\frac{P_p}{RT_p}\right) \tag{17.19}$$

$$f = 0.046\left(\frac{\overline{G}_p d_p}{\mu_p}\right)^{-0.2} \tag{17.20}$$

Here, $\overline{G_P}$ is mass velocity (kg/hr/m^2), k_p is thermal conductivity and μ_p is viscosity of permeate gas. f is the tube side friction factor calculated using Blasius correlation (17.19) applicable for turbulent flow through a circular tube.

17.4 Simulation and Optimization of MSR Operations

Total volumetric catalyst holdup V_c in the reactor tube without accounting for the volume of permeate tube is $V_c = \frac{\pi d_i^2}{4} L$, where $d_i = 79.5$ mm. Area of palladium membrane per tube is $A_m = 2\pi r_o L$. In this work, we want to study the effect of increasing palladium membrane area A_m on the overall conversion of methane, keeping the total volume of catalyst V_c in the bed constant. This is achieved by varying the inner diameter of the outer annular tube D_i as γ times d_i, i.e., $D_i = \gamma d_i$. Correspondingly, the radius r_o of the inner permeate tube is varied as $r_o = \frac{d_i}{2}\sqrt{\gamma^2 - 1}$. Thus, by increasing the parameter γ, area of the palladium membrane in the MSR is increased with total catalyst holdup volume kept constant.

Model equations derived in Sect. 17.3 are solved to simulate the performance of MSR for various operating conditions. Simulation is carried out using MATLAB® R2015a. Function subroutine *ode15 s* is used to solve 11 numbers of simultaneous stiff ordinary differential Eqs. (17.3), (17.8), (17.9), (17.10), (17.11), (17.12), (17.14) and (17.17) to obtain the profiles of mole fractions, temperature and pressure along the reactor length for various feed and operating conditions. Main factors that influence the performance of MSR operation are feed gas temperature T_i, feed gas pressure P_i, furnace gas temperature T_g, steam to methane ratio in feed S/C, hydrogen to methane ratio in feed H/C, permeate feed temperature T_{pi}, permeate feed pressure P_{pi} and sweep ratio S_r which is the ratio of feed rate of sweep gas (steam) to the feed rate of methane.

An attempt is made in this work to optimize the performance of MSR for achieving maximum methane conversion (M_c) for a fixed methane molar feed rate of $F_{CH_4} =$ 1000 kmol/h and fixed values of D/C = 0.091 and N/C = 0.02. Optimal values of operating conditions are calculated subject to certain operational constraints and bounds. Based on the creep limit on alloy steel tubes to avoid rupture, an upper bound on outer wall temperature is set as $T_{wo} \leq 1200K$ [1]. An upper bound on the total molar gas feed rate of $F \leq 5000$ kmol/h is fixed to avoid excessive pressure drop in the annular channel. Here,

$$F = \left(1 + \left(\frac{S}{C}\right) + \left(\frac{H}{C}\right) + \left(\frac{D}{C}\right) + \left(\frac{N}{C}\right)\right) F_{CH_4} \tag{17.21}$$

Following operational constraints and bounds, reported in Rajesh et al. [1] for industrial scale catalytic steam reformer are used in this work.

$$725 \leq T_i \leq 900\,\text{K}$$

$$2400 \leq P_i \leq 3000\,\text{kPa}$$

$$2.0 \leq S/C \leq 6.0$$

$$0.01 \leq H/C \leq 0.5$$

$$1375 \leq T_g \leq 1650\,\text{K}$$

Constraints and bounds set on other operating variables are as follows

$$1.0 \leq S_r \leq 3.0$$

$$600 \leq T_{pi} \leq 800\,\text{K}$$

$$200 \leq P_{pi} \leq 300\,\text{kPa}$$

The constrained optimization problem formulated in this work is solved using the MATLAB optimization routine *fmincon*. The palladium membrane area A_m is varied by changing the value of γ. Optimal values of feed and operating conditions are calculated for different values of γ in the range of $1.20 \leq \gamma \leq 2.00$.

17.5 Results and Discussions

Optimum values of feed and operating parameters and the corresponding values of methane conversion calculated for different values of γ are listed in Table 17.1. A maximum of 99.9% methane conversion is obtained for $\gamma = 2$. As the value of γ is increased from 1.75 to 2.00, corresponding increase in the value of methane conversion achieved is only 0.1% (99.8–99.9%). Such an insignificant rise in methane

Table 17.1 Optimum values of operating parameters and methane conversion for different values of γ

Operating parameters	$\gamma = 1.20$	$\gamma = 1.55$	$\gamma = 1.75$	$\gamma = 2.00$
$T_g(K)$	1650.0	1649.7	1630.4	1606.7
$T_i(K)$	879.2	900.0	900.0	900.0
$P_i(kPa)$	3000.0	2990.8	2988.5	2539.7
S/C	3.879	3.768	3.863	3.647
S_r	2.991	2.999	2.971	3.000
$T_{pi}(K)$	800.0	794.1	800.0	793.1
$P_{pi}(kPa)$	200.0	200.0	200.0	200.0
$M_c(\%)$	82.4	98.1	99.8	99.9

conversion does not justify the additional palladium cost associated with increasing the value of γ beyond 1.75. So a value of $\gamma > 1.75$ is not economical.

A decreasing trend in the furnace gas temperature T_g and the feed gas pressure P_i is observed on increasing the value of γ. This implies that the cost associated with heating of furnace gas and compression of feed gas gets reduced as the area of palladium membrane is increased. A systematic cost analysis is essential for arriving at an optimal value for γ corresponding to which the advantage of excess methane conversion and reduced operating cost nullifies the additional fixed cost incurred for palladium membrane.

17.6 Conclusions

An attempt is made in this work to develop a mathematical model to represent an industrial scale operation of a membrane steam reformer (MSR). Simulation studies carried out in this work to analyse the effect of increasing Pd membrane area on the optimal performance of MSR showed that a methane conversion of up to 99.9% can be achieved in MSR. In spite of this advantage of high methane conversion, MSR technology has not found its place in industry and this is mainly due to the high cost of preparation of Pd membrane. This scenario is expected to change in future when MSR technology becomes an economically viable choice for large scale production of hydrogen. In this context, the results presented in this work is very significant for evaluating the techno-economic feasibility of MSR as an alternative to conventional catalytic steam reformer for production of hydrogen.

References

1. Rajesh, J.K., Gupta, S.K., Rangaiah, G.P., Ray, A.K.: Multiobjective optimization of steam reformer performance using genetic algorithm. Ind. Eng. Chem. Res. **39**(3), 706–717 (2000)

2. Silva, J.D., Moraes de Abreu, C.A.: Modelling and simulation in fixed-bed and fixed-bed membrane reactors for steam reforming of methane. Int. J. Hydrogen Energy **41**(27), 11660–11674 (2016)
3. Criscuoli, A., Basile, A., Drioli, E., Loiacono, O.: An economic feasibility study for water gas shift membrane reactor. J. Membr. Sci. **181**, 21–27 (2001)
4. Johannessen, E., Jordal, K.: Study of a H_2 separating membrane reactor for methane steam reforming at conditions relevant for power processes with CO_2 capture. Energy Convers. Manag. **46**, 1059–1071 (2005)
5. Xu, J., Froment, G.F.: Methane steam reforming methanation and water-gas shift: i intrinsic kinetics. Am. Instit. Chem. Eng. J. **35**(1), 88–96 (1989)

Chapter 18
Comparison of Household Energy Consumption Pattern in Residential Buildings

G. A. Sonawane and K. S. Gumaste

Abstract Energy in various forms has become the need to survive for man. In order to attain comforting lifestyle, conventional and non-conventional resources are utilized to fullest. In the household sector, energy is utilized for various purposes like thermal comfort, kitchen utilities, entertainment, lighting, personal care utilities, etc. The consumption pattern for each individual household is different. This pattern varies with the income, standard of living, size of the family, and stock and usage of functional utility appliances, etc. In this paper, energy consumption pattern of 24 sample households (classified into three categories) is compared for various functional utilities based on usage of appliances.

Keywords Energy · Household energy consumption · Energy conservation · Survey questionnaire · Classification of households

18.1 Introduction

India is a rapidly growing economy which needs energy to meet its growth objectives in a sustainable manner. The Indian economy faces significant challenges in terms of meeting its energy needs in the coming decade. According to energypedia survey (2012), India is 8.3% deficient to supply energy efficiently. Energy sector is one of the most critical components of an infrastructure that affects India's economic growth and therefore is also one of the largest industries in India. India has the fifth largest electricity generating capacity and is the sixth largest energy consumer amounting for around 3.4% of global energy consumption. India's energy demand has grown at 3.6% pa over the past 30 years. The consumption of the energy is directly proportional to the progress of manpower with ever-growing population, improvement in the living standard of the humanity, and industrialization of the developing countries.

G. A. Sonawane (✉)
Civil Environmental Engineering, Walchand College of Engineering, Sangli 416415, India
e-mail: gayatri13121@gmail.com

K. S. Gumaste
Department of Civil Engineering, Walchand College of Sangli, Sangli 416415, India

© Springer Nature Singapore Pte Ltd. 2020

V. Sivasubramanian and S. Subramanian (eds.), *Global Challenges in Energy and Environment*, Lecture Notes on Multidisciplinary Industrial Engineering, https://doi.org/10.1007/978-981-13-9213-9_18

The amount of energy we use in our homes mainly depends on the climate where we live and the types and number of energy-consuming devices we use. The number and variety of ways we use energy in homes is changing rapidly. Energy use for air conditioning has doubled. Households currently plug in more appliances and electronics at home than ever before. While refrigerators and cooking equipment have long been standard in homes, the ownership of appliances such as microwaves, clothes washers, and dryers has increased over the past 10 years. It is increasingly common for homes to use multiple televisions and computers. Additionally, the home electronics market is constantly innovating, and new products such as DVRs, game systems, and rechargeable electronic devices are becoming ever more integral to our modern lifestyle. As a result of these changes, appliances and electronics (including refrigerators) now account for nearly one-third of all energy used in homes. Natural and electricity are the most-consumed energy sources in homes. Electricity, which is used for heating and cooling, also lights our homes and runs almost all of our appliances including refrigerators, toasters, and computers.

The—Operational Energy|| is the amount of energy required to run the building over its design life and includes electrical appliances such as air-conditioners, hot water systems, refrigeration and lighting, and the fuel consumption for completion of daily household chores. The utilization of this operational energy varies from individual to individual depending on lifestyle, income, awareness, and attitude toward the same. Currently, lighting accounts for approximately 30% of total residential electricity use, followed by refrigerators, fans, electric water heaters, and TVs.

Generally, an economy is divided into three basic income groups, viz. high-yielding, moderate-yielding, and low-yielding income group. More income enhances the stock of more no. of appliances and more utilization of private vehicles for traveling on regular basis. The hike in consumption of energy leads to increase in utilization of more fossil fuels directly or indirectly increasing the carbon emissions contributing to environment. Another approach uses micro-level data that reflect individual and household behavior; this may be referred to as a microeconomic approach.

18.1.1 Literature Review

Guan et al. [1] measured the standby power for selected set of appliances. The author carried out field measurement of standby and operational power for range of electrical appliances using energy monitor 3000. Appliances were divided into two groups. The first group was electrical appliances which are not covered by Australia's Minimum Energy Performance Standards (MEPS). The second group was white goods which are included in the MEPS but are not typically operated continuously. According to the author, an electrical appliance can be in operational mode, active standby mode, passive standby mode, or off mode and measurements for the study were taken when the devices were at lowest electrical power consumption while connected to the mains, i.e., passive standby mode. It was found that different types of electrical appliances, or different brands of a given type of appliance, use significantly different amounts of operating power and standby power. For most tested appliances, standby power usage was still well above the national goal of 1 W or less. With the exception of home audio systems, the study also found that the electrical appliances were using more operating energy than standby energy [1].

Saele and Grande [2] carried out a pilot study in Norway taking 40 sample households for the study. The metering system in Norway is on hourly basis and the units charges consumed are applied maximum for the peak hours. The author set reminders for the peak hours to be known by the households using El-button on the electrical appliances. The experiment resulted in deduction of the peak hour demand for electricity reducing the rate of charges per unit consumption for the sample households. Thus, increased demand side participation in the power market improved the market performance, and expected demand response included in the market bids on equal terms with respect to power production [2].

Dash [3] analyzed the impact of income on energy consumption for high-, low- and moderate-yielding group of sample households in Bhubaneswar, Odisa. The author considered consumption of energy in the household sector in terms of usage of the electric appliances and the fuels used for running of vehicles and other domestic purposes for all the income groups. The study was concluded defining that the electrical energy consumption is directly proportional to income of the households, whereas the consumption of fuel in form of both LPG and kerosene was maximum for moderate-yielding income group. The low-income group has maximum utilization of kerosene for lighting purpose. The author has not mentioned the methodology and database clearly which demonstrate inadequate information to understand the research [3].

Tewathia [4] conducted a survey of 395 households in north east, west, central, and southern areas of Delhi on consumption of electricity with respect to factors like income of the household, size of the house, stock and usage of appliances, dwelling size, size of family, and time spent out by the family members, etc. The author has also shown varying consumption of electricity for summer, winter, and other (monsoon, spring, and autumn) seasons. The primary survey was carried out with a questionnaire for the electricity consumption of households. An inverted U-shaped nonlinear temperature electricity curve defines that the consumption for electricity was maximum in summer season for all the economic classes. The above study concludes that the consumption of electricity was maximum in summer for all the economic classes and the high-income group contributed maximum for the consumption [4].

Mohan and Sil [5] evaluated electricity consumption of 51 households in Mumbai Metropolitan Region (MMR) based on a survey collected through personal interviews and emails. A household energy performance index was formulated which compared the electricity consumption for households of equal ability. The important factors considered for formulation of index were per capita per month consumption and number of people living in the dwelling. The outcomes of HEPI showed that it can be related to the area of the household and is directly proportional. The area of the houses can then be used to limit the power consumption. The area more than 1000 sq.ft. can have HEPI of maximum 0.5. The area of the house between 500 and

1000 sq. should be up to 0.45 and that of houses less than 500 sq.ft. should have maximum HEPI of 0.4. Classifying the findings of HEPI and area of houses in the three main areas of MMR, i.e., the Mumbai area, Thane, and Navi Mumbai, found that the HEPI was almost equal in spite of great the variation in area of houses in all the three areas [5].

Papachristos [6] explored the effect of smart meter introduction, appliance efficiency, and consumer behavior on reducing electricity consumption in the Netherlands. The author has combined two perspectives: a socio-technical approach and a bottom-up simulation approach. The range of scenarios explored the interplay between efficiency, smart meter diffusion, and consumer behavior. Effect on electricity consumption suggested that further effort is required to control and reduce it. Insights from the paper suggest that future studies should disaggregate with respect to a number of factors [6].

Literature review reveals that numerous case studies and test cases were carried out in order to determine the operational energy consumption for various places in India and in other countries as well. Seasonal variation is also represented with respect to income of high-yielding, moderate-yielding, and low-yielding income groups. Also, various energy-efficient methods to reduce the peak hour demand were introduced in some of the case studies. Smart metering system and consumer behavioral study were cost-effective analyses eventually reducing the carbon emissions.

18.1.2 Energy

Energy has become the basic need for survival of man. Energy can neither be created nor be destroyed but can be transferred from one form to another form of energy. Conversion of energy from conventional and non-conventional energy resources to electrical energy is a very common practice. As per convenience, this electrical energy is used for various appliances satisfying comfort levels of human. India's electricity generation from 1950 to 1985 was very low when compared to developed nations. Since 1990, India has recorded faster growth in electricity generation. India's electricity generation has increased from 179 TW-h in 1985–1057 TW-h in 2012. Power generation by coal fired plants and non-conventional renewable energy sources (RES) has mainly contributed to the growth in the total electricity generation, whereas the contribution from natural gas, oil, and hydro plants has decreased in the last five years (2012–2017). The gross utility electricity generation (excluding imports from Bhutan) is 1236 billion kWh during the year 2016–2017 against the corresponding actual generation of 1168 billion kWh during the year 2015–2016 with 5.81% annual growth [7].

18.1.3 Residential Buildings

A building or edifice is a structure with a roof and walls standing more or less permanently in one place, such as a house or factory. Buildings come in a variety of sizes, shapes, and functions, and have been adapted throughout history for a wide number of factors, from building materials available, to weather conditions, to land prices, ground conditions, specific uses, and aesthetic reasons. Types of buildings can be classified based on their utility are as commercial, industrial, religious, residential, educational, etc.

A building should be regarded as residential building when more than half of the floor area is used for dwelling purposes. Other buildings should be regarded as non-residential. Two types of residential buildings can be distinguished—houses (ground-oriented residential buildings): comprising all types of houses (detached, semi-detached, terraced houses, houses built in a row, etc.) each dwelling of which has its own entrance directly from the ground surface—other residential buildings: comprising all residential buildings other than ground-oriented residential buildings as defined above.

18.1.4 Operational Energy

The operating energy of a building is—the amount of energy that is consumed by a building to satisfy the demand for heating, cooling, ventilation, lighting, equipment, and appliances||. In simple terms, once the building is occupied, the energy utilized by the members living in the building till the building is demolished can be termed an operational energy of a building.

18.1.5 Economic Variation

The population in India can be broadly classified into three categories:

High-income group: income is +10 lakhs per annum; *moderate-income group*: income is 2–10 lakhs per annum; and *low-income group*: income is less than 2 lakhs per annum As the income of the household is more, the standard of living increases leading to usage of more appliances to achieve comfort level. Thus, income group plays a leading role in energy consumption.

18.1.6 Energy-Efficient Devices

Appliances can account for up to 30% of your home energy use. As our reliance on appliances increases and energy prices are also on the rise, choosing energy-efficient appliances becomes more important. The national standards for energy efficiency are improving the environmental performance of appliances all the time, so upgrading to a more efficient appliance can save you energy and money. It is not only about having the right product—how you use appliances in your home can make a big difference. For example, washing your clothes with cold water can save up to five times more energy than a warm wash.

18.2 Methodology

18.2.1 Selection of Sample Households

A questionnaire survey was done for randomly chosen households based on accessibility, locality, no. of appliances, frequency of usage, and duration for which the appliances are utilized and further screening was done and 24 sample households were chosen (eight samples of each yielding group). Sample of questionnaire is attached in the Annexure 1.

18.2.2 Classification of Households Based on

18.2.2.1 Number of Appliances

The classification of list of appliances is given below. The appliances considered are grouped together as per their need to be used for defined function as functional utilities (Table 18.1).

18.2.2.2 Based on the Number of Appliances, the Households Were Classified into Three Types

HIG: Households with +70% of appliances
MIG: Households with 30–70% of appliances
LIG: Households with less than 30% of appliances.

Table 18.1 Classification of appliances into functional utilities

Kitchen appliances	1	Water purifier
	2	Electric chimney
	3	Mixer/grinder/juicer
	4	Microwave
	5	Refrigerator
	6	Exhaust fan
Entertainment	7	Music system
	8	DVD player
	9	Mobile phones or tablets
	10	Computer
	11	Laptop
	12	Television
Lighting	13	CFL
	14	LED
	15	Tube light
	16	Tungsten bulb
Thermal comfort	17	Ceiling fan
	18	Wall mounted fan
	19	Table fan
	20	Cooler
	21	A/C
Personal care utilities	22	Hair straightener
	23	Trimmer
	24	Blow dryer
	25	Iron
Others	26	Motor/pump
	27	Vacuum cleaner
	28	Washing machine
	29	Exercising unit
	30	Mini floor mill

18.2.2.3 The Income of the Households

HIG: Income is more than 10 lakh rupees per annum.
MIG: Income lies between 2 and 10 lakh rupees per annum.
LIG: Income is less than 2 lakh per annum.

18.2.3 Data Collection

Data collection was done by questionnaire survey. Also, the photograph of light bills of the households was collected for the month of September, October, November, and December.

As per the information provided in the questionnaire, the number of units consumed by each functional utility was calculated and the total was cross verified with the number of units on the light bill for the respective month.

18.2.3.1 Sample Calculation

Sample calculation is attached in the Annexure 2.

The sample calculation is the detailed representation of energy consumption based on the questionnaire survey. The total number of calculated units of energy consumed is tallied with number of units consumed by the households as per the electricity bill for the respective month.

The sample bill for the respective sample calculation is attached in Annexure 3

18.3 Results

As shown in the sample calculation, similar data collection was carried out for each of the 24 sample households and the total units consumed for each functional utility is estimated. A comparison sheet was prepared for each of the months September, October, November, December, and January with a detailed comparison with unit consumption for each functional utility (Fig. 18.1).

Sample comparison sheet for September is attached in Annexure 4.

The above graph explains the total energy consumption pattern in each of the economic classes. The drop in the consumption pattern for high-income group (HIG) with the progressing months was observed since the usage of appliances of thermal comfort like fan has reduced. The reduction in the temperature on the months of November, December, and January is noticeable. For moderate-income group (MIG), reduction in usage is sudden due to thermal comfort appliances and kitchen appliances. An increase in the lighting utility is observed for all the income groups since the natural light is not adequate enough in the months of November, December, and January. The HIG and some of the MIG sample households are aware of the energy-efficient lighting system. Whereas, LIG sample households use conventional lighting appliances leading to increase in number of units consumed due to lighting appliances (Fig. 18.2).

The average energy consumption as per the classified functional utilities is expressed in the above graph. Kitchen utilities contribute maximum since the usage

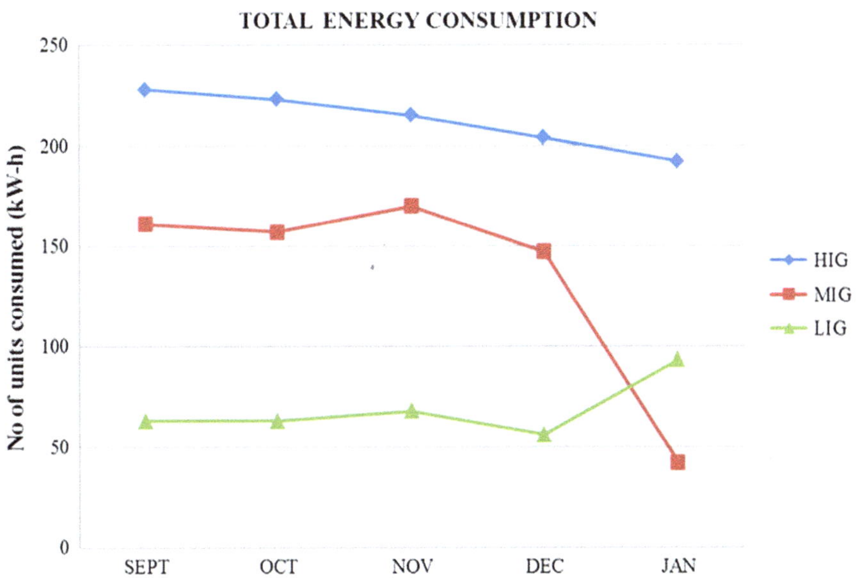

Fig. 18.1 Total average energy consumption

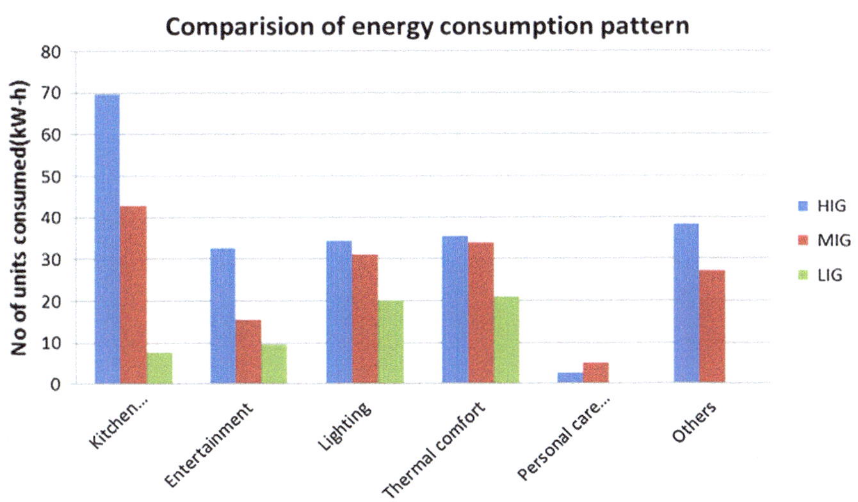

Fig. 18.2 Comparison of average energy consumption for functional utilities

of appliances consuming power like microwave oven, electric chimney, and juicer refrigerator is more frequent for more duration compared to MIG and LIG.

The usage of the personal care utilities is more in MIG since the sample households cannot afford frequent visits to expensive salons or laundry expenses like HIG samples. The usage of self care or grooming appliances like Hair Straightner (Fig. 18.3).

The above graph shows energy consumption for HIG. Kitchen appliances like microwave oven, mixer juicer, and electric chimney are used more frequently in HIG and thus kitchen appliances have maximum contribution. The contribution from other appliances varies with personal usage habits of sample households (Fig. 18.4).

The personal utility consumption is maximum in case of MIG since this group cannot afford visiting expensive salons frequently to groom themselves. The entertainment appliances consumption is more contributed due to usage of television by housewives compared to HIG (Fig. 18.5).

The number of units consumed in personal utilities and others for LIG is observed to be negligible and zero, respectively. The reason for this is low-income group cannot afford appliances falling in this category.

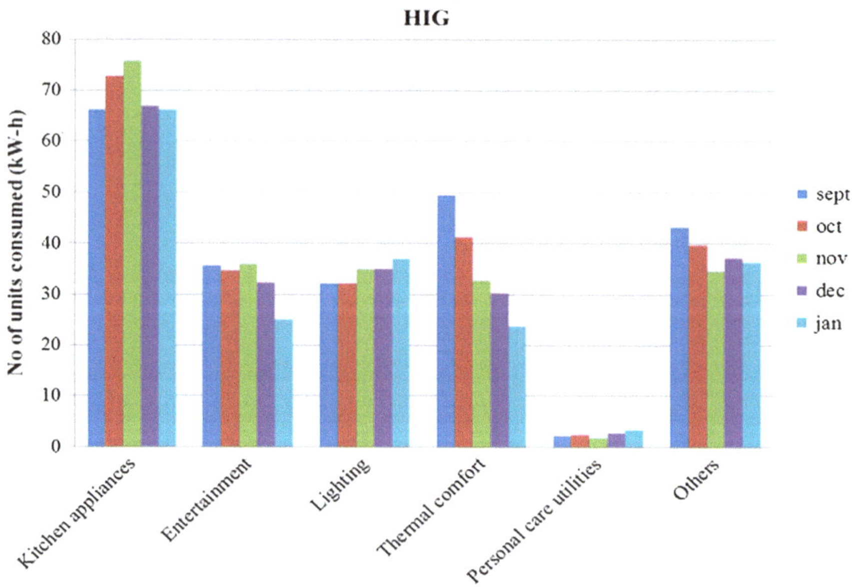

Fig. 18.3 Energy consumption pattern for high-income group

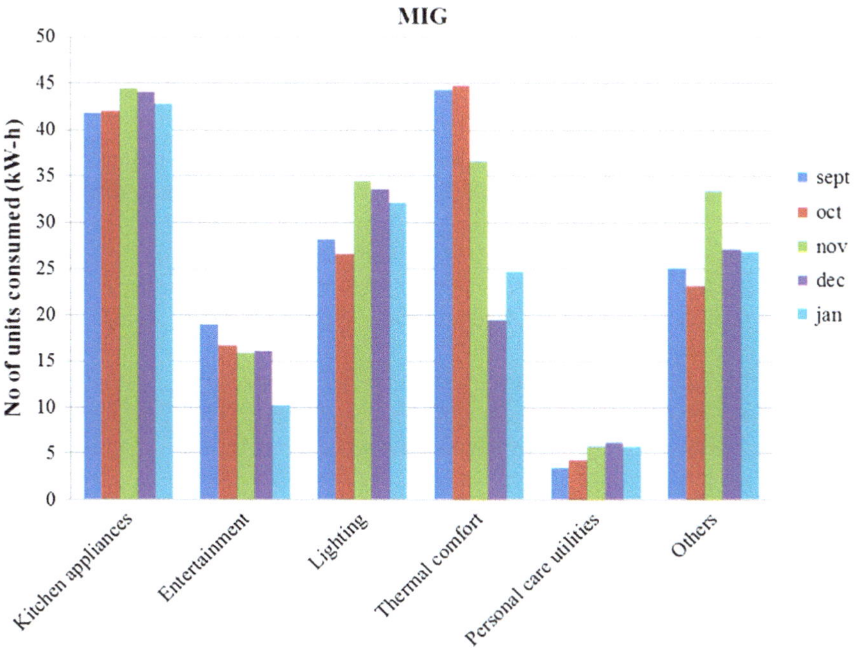

Fig. 18.4 Energy consumption pattern for moderate-income group

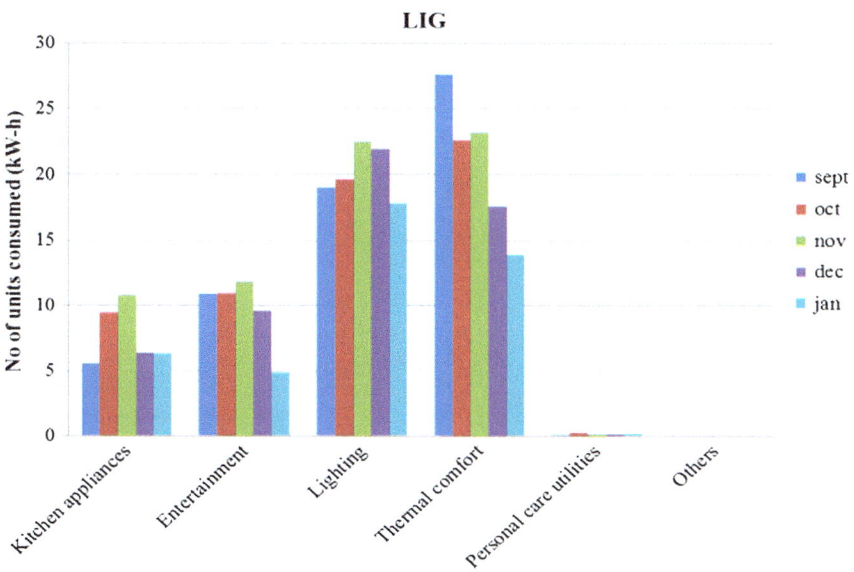

Fig. 18.5 Energy consumption pattern for low-income group

18.4 Conclusion

1. The energy consumption for high-income group is more than moderate-income group and low-income group.
2. Pattern of consumption showed that thermal comfort appliances contribute more to the consumption for MIG and LIG compared to HIG due to unawareness regarding energy-efficient appliances. Also, the construction of HIG sample households is done considering factors like natural lighting compared to MIG and LIG. As the durability and energy efficiency of lighting appliances increase, the cost of the appliances is more, which is not affordable for all MIG samples and LIG samples.
3. Other appliances contribution varies in each of the classes based on the standard of living, habits, number of members in the family, area of the sample household, and frequency of usage of sample as per the need.

Acknowledgements I would like to express gratitude toward TEQIP phase 3. I would like to thank all the sample household owners for the cooperation. I would also like to thank my friends and family for the direct or indirect support.

Annexures

See Annexure Tables 18.2 and 18.3.

Annexure
See Annexures Fig. 18.6.
See Annexure Table 18.4.

18.2 Annexure I

								Usage				
		Yes/no	Quantity	Used regularly	Capacity	Brand	Watt	Daily	Weekly	Monthly	Watt-h	Total
Background info	Name:											
	No of members in the house											
	What is the occupation of family members											
	What r the working hours											
	Income											
	No of rooms											
	Built-up area											
	Monthly light bill											
Stock of appliances	*Kitchen appliances*											
	Water purifier											
	Electric chimney											
	Mixer/grinder/juicer											
	Microwave											
	Refrigerator											
	Exhaust fan											
	Entertainment											
	Music system											

(continued)

18.2 (continued)

							Usage				
	Yes/no	Quantity	Used regularly	Capacity	Brand	Watt	Daily	Weekly	Monthly	Watt-h	Total
DVD player											
Mobile phones or											
Tablets											
Computer											
Laptop											
Television											
Lighting											
CFL											
LED											
Tubelight											
Tungstun bulb											
Thermal comfort											
Ceiling fan											
Table fan											
Wall mounted fan											
Cooler											
Room heater											
A/C											

(continued)

18.2 (continued)

							Usage				
	Yes/no	Quantity	Used regularly	Capacity	Brand	Watt	Daily	Weekly	Monthly	Watt-h	Total
Personal care utilities											
Hair straightner											
Trimmer											
Blow dryer											
Iron											
Others											
Motor/pump											
Vacuum cleaner											
Washing machine											
Exercising unit											
Total (kWh)											
No of units in bill (kWh)											

18.3 Annexure II

									Usage				
			Yes/no	Quantity	Used regularly	Capacity	Brand	Watt	Daily	Weekly	Monthly	Watt-hr	Total
Background info	Name: Shailendra Kelkar												
	No of members in the house	4											
	What is the occupation of family members	Buisness											
	What r the working hours	10											
	Income												
	No of rooms	4 Bhk											
	Built-up area	2500											
	Monthly light bill	1800–2000											

(continued)

18.3 (continued)

Stock of appliances	*Kitchen appliances*												
	Water purifier		Y	1	1	8 L	Kent	60	30 min		15	900	104.9
	Electric chimney		Y	1	1		Craft	250				0	
	Mixer/grinder/juicer		Y	1	1	3 Jars	Philips	600	20 min		10	6000	
	Microwave		Y	1	1	23 L	LG	800	20		10	8000	
	Refrigerator		Y	1	1	321 L	LG	1000				30,000	
	Exhaust fan		Y	3	2		Havells	500	2		60	60,000	
	Entertainment											0	
	Music system		Y	1	1		Sony	60	0		0	0	15.05
	DVD player		Y	1	1		Videocon	240			0	0	
	Mobile phones or tablets		Y	4	1		iphone	4	1		30	120	
					1		iphone	4	1		30	120	
					1		iphone	4	1		30	120	
					1		iphone	4	1		30	120	
	Computer		y	1	1		hp	250		2	8	2000	
	Laptop		y	2	1		hp	60	2		60	3600	
					1		Dell	60			2	120	

(continued)

18.3 (continued)

	Television		Y	1	1	42″	Videocon	80	3		90	7200	
				1	1	32″	Sony	55	1		30	1650	
	Lighting											0	
	CFL		Y	54	8		Urja	6	7		210	10,080	11.34
				5	3		Philips	4	2		30	360	
	LED		Y	25	6		Philips	5	1		30	900	
	Tubelight		N										
	Thermal comfort												
	Fan		Y	5	2	4 blades	Bajaj	25	6		180	9000	9
	Table fan		Y	5	0		Havells	25	0		0	0	
	Wall mounted fan		Y	4			Havells	20			0	0	
	Cooler		N									0	
	A/C		Y	4	4	1.5 tonne	Samsung				0		
	Personal care utilities												
	Hair straightner		Y	2	2		Sony	280		15 MIN	1	560	2.23
	Trimmer		Y	1	1		Philips	40			10	400	
	Blow dryer		Y	1	1		Nova	70		15 MIN	1	70	
	Iron		Y	1	1		Kenstar	600		30 min	2	1200	
	Others											0	

(continued)

18.3 (continued)

	Motor/pump		Y	1	1	0.5 HP	Kirloskar	373	10 MIN		5	1865	33.195
				1	1	1 HP	Kirlo	746	10		5	3730	
							Skar						
	Vaccum cleaner		Y	1	1		Eureka forbes	800		3	12	9600	
	Washing machine		Y	1	1	6.5 l	LG	600	60 MIN		30	18,000	
	Exercising unit		Y	1	1		Matrix	300			0		
	Total (kWh)												175.72
	No of units in bill (kWh)												175

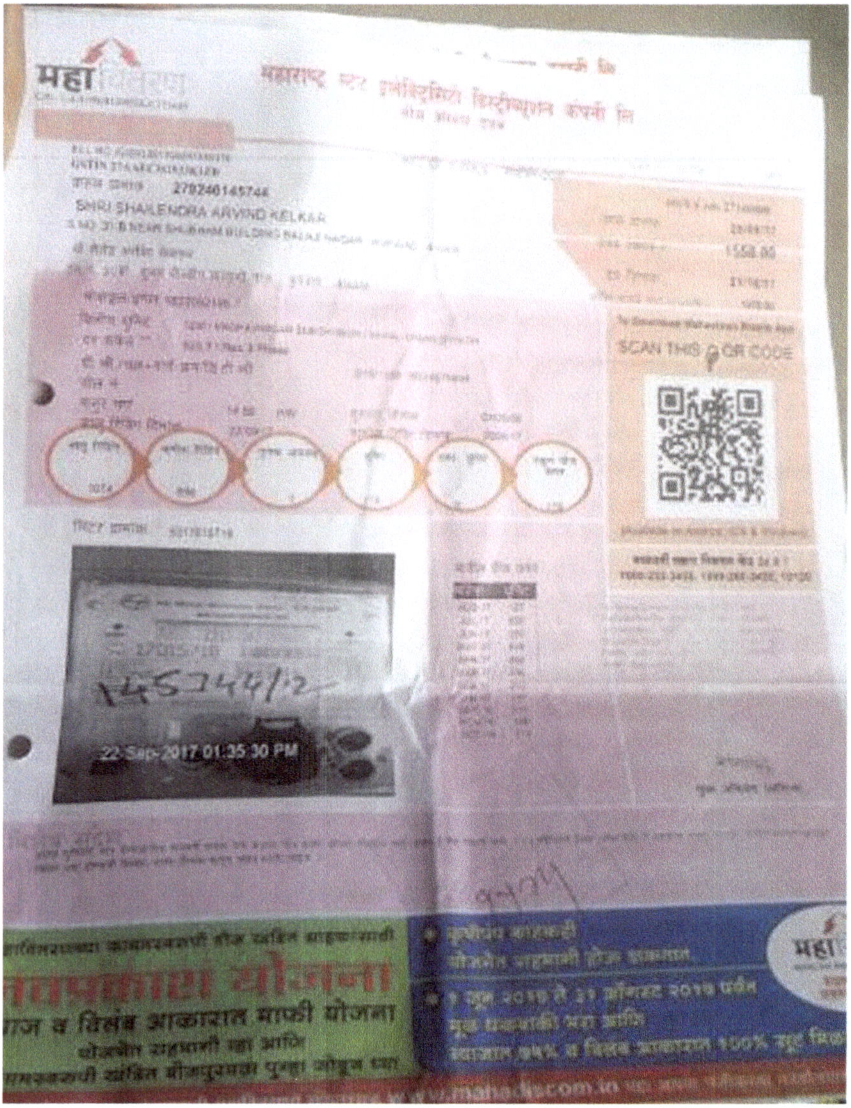

Fig. 18.6 Sample electricity bill

18.4 Annexure IV

Sr. no	Income group	Name of household	No of members	Built-up area	Kitchen appliances	Entertainment	Lighting	Thermal comfort	Personal care utilities	Others	Total
1	HIG	Addappa Mankapure	5	1200	63.45	10.44	0.216	18	2.001	7.33	
2		Gayatri Sulakhe	3	2500	47	24.96	40.2	58.5	2.95	47.73	
3		Nilkanth Bijargi	3	2200	57.35	76.26	90.78	81	3.88	70.255	
4		Rajkumar Herale	2	1400	31.75	4.56	6.24	4.5	0	23.665	
5		Shailendra Kelkar	4	2500	101.9	15.05	12.06	13.5	2.03	33.195	
6		Shashank Patil	3	4000	52.35	32.76	25.29	67.5	4.6	28.46	
7		S. V. Kulkarni	5	3000	110.6	53.58	41.46	100.8	0	61.66	
8		Zakhir Patait	5	4000	64.9	67.56	40.8	51.9	1.88	73.19	
		Avg			66.16	35.64	32.13	49.46	2.167	43.18	228.75
		%			28.92	15.58	14.04	21.62	0.947	18.87	
		Standard deviation			26.940	27.09	28.79	34.53	1.646	23.83	
1	MIG	Annasaheb Manohar Rote	2	2000	36.4	10.08	55.26	24	4.8	34.86	

(continued)

18.4 (continued)

Sr. no	Income group	Name of household	No of members	Built-up area	Kitchen appliances	Entertainment	Lighting	Thermal comfort	Personal care utilities	Others	Total
2		Madan Joshi	5	750	36.18	21.96	33.72	54	4.8	29.745	
3		Mangala Kedge	3	2200	40.35	8.34	33.84	36	10.64	20.4	
4		Satish Patil	2	1300	63.9	18.24	15.06	10.5	4.8	37.185	
5		Sulochana Shinde	6	1500	39.9	32.52	31.38	105	1.2	25.372	
6		Mali	2	1100	49.8	21.02	24.06	50.4	0	34.79	
7		Vijay Chaugule	4	680	35.7	19.44	18	36	1.2	18	
8		Yellapa Khobra	8	600	31.84	20.16	14.22	37.8	0	0	
		Avg			41.75875	18.97	28.193	44.2125	3.43	25.044	161.608
		%			25.839571	11.7383	17.445	27.3579	2.12242297	15.49678	
		Standard deviation			10.387251	7.473228	13.596	28.1585	3.60734172	12.30178	
1	LIG	Chandar More	13	320	2.4	26.7	50.4	81	0	0	
2		Dhondiran Londe	3	100	4	0.12	19.2	31.5	0	0	
3		Ganesh Shinde	4	150	2.4	4.8	5.52	16.8	0	0	

(continued)

18.4 (continued)

Sr. no	Income group	Name of household	No of members	Built-up area	Kitchen appliances	Entertainment	Lighting	Thermal comfort	Personal care utilities	Others	Total
4		Godabai Dalwai	11	500	30	23.1	42.18	31.5	0	0	
5		Hanuman Patil	3	120	0.6	3.3	2.4	13.5	0.6	0	
6		Maruti Pujari	9	350	4.8	26.52	25.92	36	0	0	
7		Rajaram Biranje	8	150	0	2.416	1.8	0	0	0	
8		Sarode	4	120	0.115	0	4.14	10.5	0	0	
		Avg			5.539375	10.8695	18.945	27.6	0.075	0	63.0289
		%			8.7886306	17.24527	30.058	43.7895	0.11899308	0	
		Standard deviation			10.038895	12.21488	19.068	24.8486	0.21213203	0	

References

1. Guan, Lisa, Berrill, Trevor, Brown, Richard J.: Measurement of standby power for selected electrical appliances in Australia. Energy Build. **43**, 485–490 (2010)
2. Saele, H., & Grande, O.S.: Demand response from household customers: experiences from a pilot study in Norway. IEEE Trans. Smart Grid, **2**(1), 102–109 (2011)
3. Dash, M.: An economic analysis of household energy consumption of urban Odisha, Int. J. Sci. Res. (IJSR) **4**(7) (2015)
4. Tewathia, N: Determinants of the household electricity consumption: a case study of Delhi. Int. J. Energy Econ. Policy **4**(3), 337–348 (2014)
5. Mohan, S., Sil, A.: Evaluation of electricity consumption and development of household energy performance index (EPI): a case study of mumbai metropolitan region (MMR). J. Energy Res. Environ. Technol. **1**(1), 14–19 (2013)
6. Papachristos, G.: Household electricity consumption and CO_2 emissions in the Netherlands: a model-based analysis. Energy Build. **86**, 403–414 (2014)
7. Kumar, S., et al.: Present and future energy scenario in India. (2013). https://doi.org/10.1007/s40031-014-0099-7

Chapter 19
Classical PID Controller for Semi-active Vibration Control of Seismically Excited Structure Using Magneto-Rheological Damper

Kavyashree, H. M. Jagadisha, Vidya S. Rao and Bhagyashree

Abstract In this paper, a Proportional Integral Derivative (PID) controller is designed to mitigate the vibration response of the structure. The simulation is done for three degrees of freedom system under different earthquakes using MATLAB and Simulink. Different earthquakes are considered to find the capability of the controller used. The controller is designed using tuning rules of Zeigler–Nichols ultimate gain method. The controller proposed will reduce the vibrations of the structure as compared to the vibrations obtained without controller. The controller output is given to Magneto-rheological (MR) damper which gives necessary force output, which will be applied to the structure to damp out its vibrations.

Keywords Magneto-rheological damper · Proportional integral derivative controller · Semi-active control

19.1 Introduction

Earthquake is one of the most severe natural hazards that cause damage to structures. Recent damaging earthquakes provide powerful reminders of how our built environment is vulnerable to the forces of nature. Consequently, one of the principal challenges in structural engineering concerns the development of innovative concepts to better protect structures, along with their occupants and contents, from

Kavyashree (✉) · H. M. Jagadisha · V. S. Rao · Bhagyashree
Manipal Institute of Technology, Manipal Academy of Higher Education, Manipal, India
e-mail: kavya.shree@learner.manipal.edu

H. M. Jagadisha
e-mail: jagadisha.hm@manipal.edu

V. S. Rao
e-mail: rao.vidya@manipal.edu

Bhagyashree
e-mail: bhagya.shree@manipal.edu

© Springer Nature Singapore Pte Ltd. 2020

V. Sivasubramanian and S. Subramanian (eds.), *Global Challenges in Energy and Environment*, Lecture Notes on Multidisciplinary Industrial Engineering, https://doi.org/10.1007/978-981-13-9213-9_19

the damaging effects of environmental forces including those due to earthquakes, cyclones, and tsunamis.

Indian subcontinent has severe earthquake threat and the past two decades have seen devastating earthquakes striking India with frightening regularity. Most of these buildings are not earthquake resistant and are potentially vulnerable to collapse in the event of a high-intensity earthquake. Hence, development of controllers for seismic response control and damage mitigation technology of structures needs to be elaborated. From the last few decades, the concern about the stability of the structure has gained importance because the safety of the dwellers depends on the stability of the structure. The structural protective system like passive, active and semi-active control have been adopted to mitigate the vibration of the structure which are stated in [1]. Nowadays, the design of the structures without the control system is considered as threat to the dwellers of the structures. These structures need a sophisticated controller design technique to resist earthquakes. From the fast literature it is evident that clipped optimal control algorithm was used to reduce response of the system with feedback of the structural acceleration which showed the reduction in the structural response found in [2]. By using the active mass driver system with the feedback of the structural response, there was reduction in the structural response in [3]. All these controller-commanded voltages have to be converted to force, which is produced by dampers. Different models of magneto-rheological dampers were proposed in [4]. Lyanpunov controller, modulated homogeneous friction algorithm, and a decentralized bang–bang controller with the MR damper was used to find the efficiency of the controllers used in [5]. The placement of the MR damper in different position by varying the natural frequency and damping of the structure also affects the response of the structure in [6]. Proportional integral derivative controller and a non-chattering robust sliding mode controller were used separately to reduce the response of the structure excited for Marmara earthquake of Turkey in 1999 of magnitude 7.4 for a multistoried building [7]. Clipped optimal control algorithm adapted to control MR damper in controlling the bench mark three story structure, where the controller was compared with passive on and off system for the performance ability is presented in [8]. The protective system with the control algoritm is also adopted for base isolation system to mitigate the vibration of the structure. $H\infty$ optimal feedback algorithm which utilizes a fuzzy logic approach to establish weighting functions is used with the passive, active and semiactive dampers to compare the ability of the dampers in reduction of response of the structure with base isolation is presented in [9]. From the previous literature survey done, it indicates that there is a need to improve the semi-active protective system with suitable control algorithm in controlling the structure excited for earthquake. Therefore this research aims at controlling a three story benchmark structure excited for the earthquake force in single direction by using MR damper with current driver which is controlled by PID control algorithm.

19.2 Magneto-rheological Fluid Dampers

19.2.1 Modeling and Performance of MR Damper

The property of the magneto-rheological fluid is attractive and this reason makes it promising material to be used in the dissipation of force when used in the magneto-rheological damper. The damper has a behavior of stable hysteresis for a long range of temperature from −40 to 150 °C, its has high yielding strength, it also has large force output, low power requirment, high dynamic range and low viscosity which makes it promising in the structural control as stated in [4, 10].

These fluids contain magnetically polarizable particles which are of micron size which are dispersed in a medium like silicon or mineral oil. For this fluid, when a magnetic field is passed the particles, they form a chain and this fluid changes to semi-solid providing the plastic property. In a few seconds, the transition to the rheological equilibrium can be produced, and an experiment was conducted to find the capacity of MR damper in the application of the structural response reduction. They developed accurately a prototype model of MR damper to take the advantage of its unique feature [4].

The simple mechanical idealizations of the MR damper in Fig. 19.1 have been shown to accurately predict the behavior of the prototype MR damper over a broad range of inputs. The force predicted by MR damper model is given in the Eq. 19.1.

$$f = \alpha z + c_o(\dot{x} - \dot{y}) + k_o(x - y) + k_1(x - x_o) \tag{19.1}$$

$$\dot{z} = -\gamma|\dot{x} - \dot{y}|z|z|^{n-1} - \beta(\dot{x} - \dot{y})|z|^n + A(\dot{x} - \dot{y}) \tag{19.2}$$

$$\dot{y} = \frac{1}{(c_o + c_1)} \times \{\alpha z \dot{+} c_o\dot{x} + k_o(x - y)\} \tag{19.3}$$

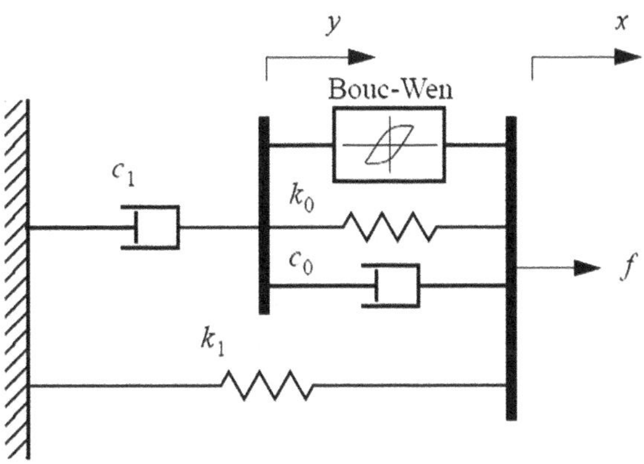

Fig. 19.1 Mechanical model of MR damper [4]

Here, k_1 represents the accumulator stiffness; c_o represents the viscous damping at the large velocity. k_o controls the stiffness when there are large velocity, x_o represents initial displacement of spring associated with the nominal damper force due to the accumulator. c_1 represents nonlinearity in the force–velocity loop and z is the evolutionary variable. The shape of the hysteresis loop is controlled by varying the parameters A,γ, and β to account for the dependence of the force on the voltage applied to the current driver (a detialed mechanism of current driver is elucidated in [11]) and the resulting magnetic current, [4] have suggested,

$$\alpha = \alpha(u) = \alpha_a + \alpha_b u \tag{19.4}$$

$$c_1 = c_1(u) = c_{1a} + c_{1b} u \tag{19.5}$$

$$c_o = c_o(u) = c_{oa} + c_{ob} u \tag{19.6}$$

$$\dot{\mathrm{u}} = -\eta(\mathrm{u} - \mathrm{v}) \tag{19.7}$$

where output of the controller is u and the voltage commanded sent by the controller is v. The details of parameters are given [4].

19.3 Proportional Integral Derivative Controller Design

The widely used controller in the industries is the PID controller. There are proportional (P) controller, proportional derivative (PD), and proportional integral derivative (PID). In this study, the PID controller is designed. This controller works on the calculated error in a closed-loop system. The PID is tuned according to Ziegler–Nichols tuning rule. The ultimate gain method is used to tune the PID controller. Here, $Y(t)_{\text{ref}}$ represents the desired value for the output of the system and $y(t)$ represents output of the system. Here, $e(t)$ represents error calculated and $u(t)$ represents the controlled signal. MATLAB and Simulink are used in this study for the control process and the simulation. The governing equation of the PID controller is given in Eq. 19.8 which is presented in [12].

$$u(t) = K_p \times \left\{ e(t) + \left(\frac{1}{T_i} \times \int_0^t e(t) \times d(t) \right) + \left(T_d \times \frac{\mathrm{d}e(t)}{\mathrm{d}t} \right) \right\} \tag{19.8}$$

where K_p represents proportionality constant, T_i represents integral time, and T_d represents derivative time. Ziegler–Nichols ultimate gain rule is used to tune the PID controller. Figure 19.2 shows the general closed-loop feedback controller system [12].

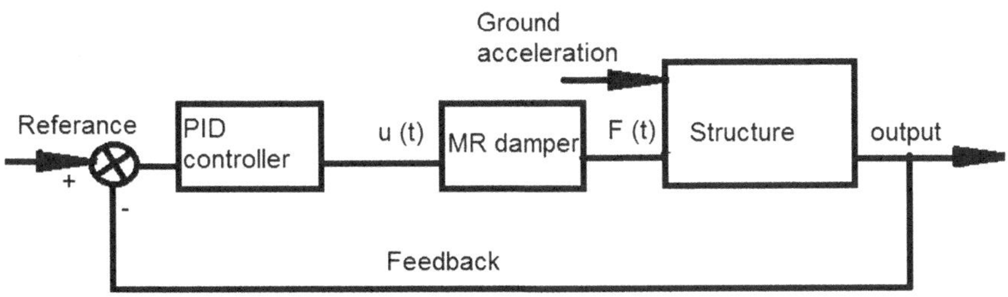

Fig. 19.2 Closed-loop block diagram

19.4 Numerical Problem

The three-story benchmark building is taken from the literature [13]; the structure is exited for El-Centro earthquake 1940. The structural system is converted to state space as shown in the equations below,

$$\dot{x}(t) = Ax(t) + Bu(t) \tag{19.9}$$

$$y(t) = Cx(t) + Du(t) \tag{19.10}$$

The input is taken as the ground acceleration and the damper forces indicated as m, the state vectors are taken as velocity and the displacement of the three-story indicated as n, the output is taken as velocity of the structure indicated by p, A is $n \times n$ system matrix, B is $n \times m$ input matrix, C is $p \times n$ output matrix, and D is $p \times m$ feed-forward matrix.

The structural parameters are taken from the benchmark building and equated to the structural motion equation stated by Chopra [14].

$$M\ddot{U} + C\dot{U} + KU = RF_u - M\ddot{U}_g \tag{19.11}$$

Here M, C, and K represent the mass, damping, and stiffness of the structure. U, $\dot{U}$, and $\ddot{U}$ represent displacement, velocity, and acceleration of the structure. R represents the location of the MR damper, F_u represents the force of the MR damper, and $\ddot{U}_g$ represents the ground acceleration which are obtained from previously recorded event presented in [15]. This equation of the structural motion is converted to the state space has shown above. The structural parameters are considered from [13]. By using MATLAB and Simulink, the simulation is carried.

19.5 Results and Discussion

The MR damper rigged up in the Simulink block is given the input as the constant voltage and the displacement of sine wave with frequency of 2.5 Hz and amplitude of 1.5 cm. The voltages given are 0, 0.75, 1.5, and 2.25 V. The graph is plotted for the MR damper force versus time, force versus displacement, and force versus velocity. The graph plotted is comparable with the graphs obtained in [4].

The condition of controlled and uncontrolled forces of the structure subjected to the earthquake is modeled in the MATLAB and Simulink. The force produced by the damper, displacements, and the acceleration of the three floors of the building excited for the El-Centro of 1940 earthquake when the controller and damper are placed in between ground and first floor are shown in Figs. 19.3 and 19.4, respectively.

The graphs obtained above show the force predicted by the MR damper to reduce the vibration of the structure. The displacement versus time graph shows the response of the structure in the absence of the controller and in the presence of the controller. The graphs show uncontrolled and the reduced response of the structure in the absence and presence of the PID controller. The acceleration versus time graphs also show the uncontrolled and the controlled response of the structure in the absence and presence of the PID controller.

The above-obtained values of the displacement and acceleration are compared with the value obtained in Dyke et al. [2]. The uncontrolled values are in close agreement with the results obtained in the Dyke et al. [2]. The PID-controlled results give a much reduced response than uncontrolled response of the structure. This result shows that the PID controller used to control response of the structure is capable in controlling the structural vibration. Therefore, the structure excited for earthquake can be controlled using the PID controller with MR damper when placed in between ground and first floor in the considered benchmark structure Table 19.1.

When the damper is placed in between the ground and the first floor, there is a reduction in the displacement for about 89.6% in the first floor, 90.2% in the second floor, and 88% of reduction in the third floor. Acceleration is reduced for about 77.9% in the first floor, 86.97% in the second floor, and 83.1% of reduction in the third floor. The force required by the MR damper to sustain the vibration is 807.89 N. These values of the uncontrolled parameters are similar to the Dyke et al. [2] where they have used the clipped optimal control and here the controller is proportional integral derivative.

Simulation is conducted in a similar manner for the Northridge and Kobe earthquake. The result shows a reduction in the displacement and also acceleration of all the stories. The peak force produced for the three different earthquakes considered in the study is mentioned in the Table 19.2, which conclude that 3000 N capacity of the damper used can be placed in between ground and first floor for the three different earthquakes considered.

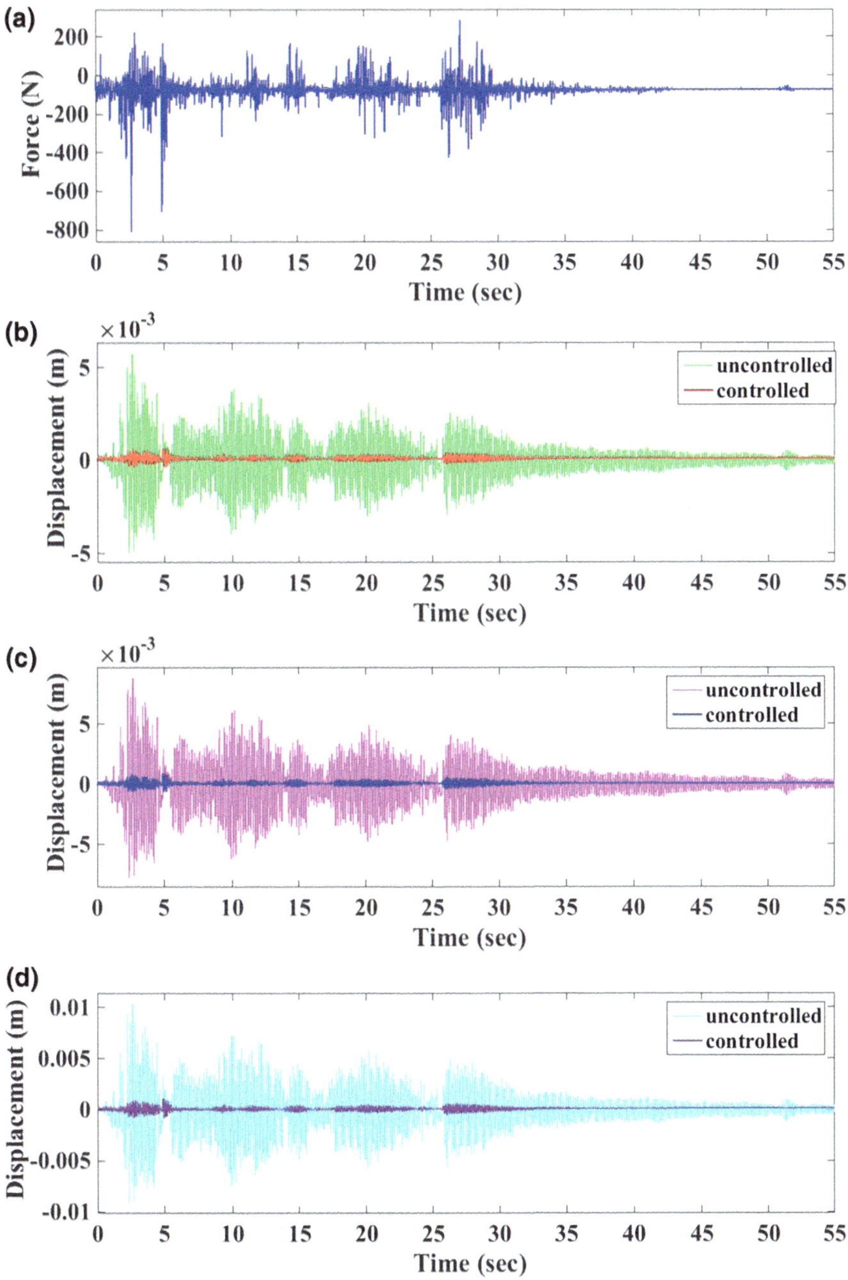

Fig. 19.3 **a** Force versus time graph produced by MR damper, **b** displacement versus time graph of the first floor, **c** displacement versus time graph of the second, and **d** displacement versus time graph of the third floor, when damper is placed in between the ground floor and first floor

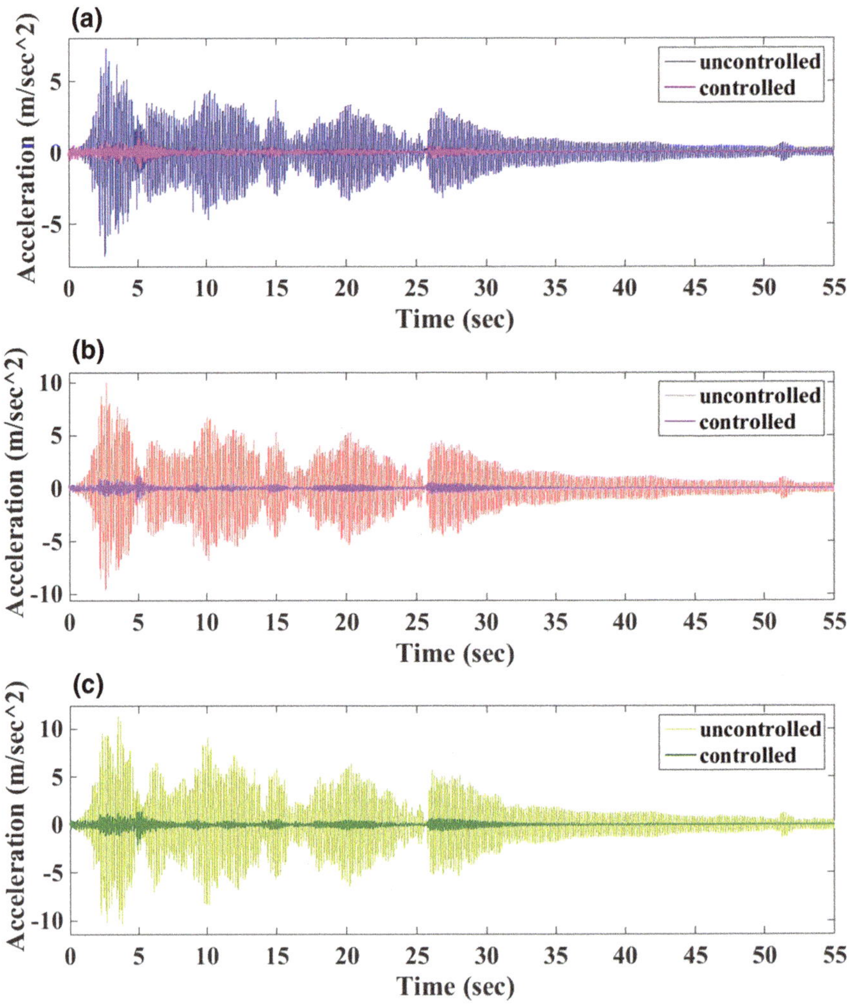

Fig. 19.4 **a** Acceleration versus time graph of the first floor, **b** acceleration versus time graph of the second, and **c** acceleration versus time graph of the third floor, when damper is placed in between the ground floor and first floor

19.6 Conclusion

19.6.1 Conclusion of the Work

Vibration of the structure due to the excitation of earthquake is mitigated by using the PID controller. This PID controller is designed by a classical method using Ziegler–Nichols tuning rule by the ultimate gain method. This designed controller

Table 19.1 Peak response of the structure due to El-Centro earthquake when damper is placed in between ground and first floor

Control strategy	Story number	Uncontrolled	PID-controlled	Peak reduction in (%)
Displacement (cm)	U_1	0.60	0.0624	89.6
	U_2	0.93	0.091	90.2
	U_3	1.00	0.12	88
Acceleration (cm/s^2)	$\ddot{U}_1$	724	159.97	77.9
	$\ddot{U}_2$	993.71	129.4	86.97
	$\ddot{U}_3$	1242.58	209.9	83.1
Force (N)	F_u	0	807.89	–

Table 19.2 Force produced by the MR damper when placed in between the ground and the first floor for the three earthquakes considered

Earthquakes	Force produced by MR damper (N)
El-Centro	807.89
Northridge	1328
Kobe	202.3

will reduce the vibrations of the structure when compared to the uncontrolled condition. This controller has shown that it is efficient in controlling the structural response for the three different earthquakes considered in the study. This voltage produced is commanded to the MR damper through current driver which adequately performs with the controller and produces the required amount of force to sustain the vibration. The MR damper of 3000 N capacity can be used to sustain the vibration when the damper is placed in between the ground and the first floor for the three different earthquakes considered in the study.

For the El-Centro earthquake, when damper is placed in between ground and first floor, the response of the structure reduces with the less amount of force produced by the actuator. In Northridge and Kobe earthquake, the PID controller proposed has controlled the structure with the minimum amount of force produced by the MR damper. Therefore, the PID controller which is designed with the MR damper can be used in the structure response control.

References

1. Housner, G.W., Bergman, L.A., Caughey, T.K., Chassiakos, A.G., Claus, R.O., Masri, S.F., Skelton, R.E., Soong, T. T., Spencer Jr., B.F., Yao, J.T.P.: Structural control: past, present, and future. J. Eng. Mech. **123**(9), 897–971 (1997)
2. Dyke, S.J., Spencer Jr., B.F., Sain, M.K., Carlson, J.D.: Modeling and control of magnetorheological dampers for seismic response reduction. Smart Mater. Struct. **5**(5), 565–575 (1996)
3. Dyke, S.J., Spencer Jr, B.F., Quast, P., Kaspari Jr, D.C., Sain, M.K.: Implementation of an active mass driver using acceleration feedback control. Comput.-Aid. Civil Infrastru. Eng. **11**(5), 305–323 (1996)

4. Spencer Jr., B.F., Dyke, S., Sain, M.: Phenomenological model for magnetorheological dampers. J. Eng. **123**(3), 230–238 (1997)
5. Jansen, L.M., Dyke, S.J.: Semiactive control strategies for MR dampers: comparative study. J. Eng. Mech. **126**(8), 795–803 (2000)
6. Villarreal, K., Wilson, C., Abdullah, M.: Effects of MR damper placement on structure vibration parameters. In: Proceedings of the 2004 Earthquake Engineering Symposium for Young Researchers (2004)
7. Guclu, R.: Sliding mode and PID control of a structural system against earthquake. Mathe.Comput. Model. **44**(1–2), 210–217 (2006)
8. Dyke, S.J., Spencer Jr., B.F., Sain, M.K., Carlson, J.D.: An experimental study of MR dampers for seismic protection. Smart Mater. Struct. **7**, 693–703 (1998)
9. Wongprasert, N., Symans, M.D.: Numerical evaluation of adaptive base-isolated structures subjected to earthquake ground motions. J. Eng. Mech. **131**(2), 109–119 (2005)
10. Yang, G., Spencer Jr., B.F., Carlson, J.D., Sain, M.K.: Large-scale MR fluid dampers: modeling and dynamic performance considerations. Eng. Struct. **24**(3), 309–323 (2002)
11. Yang, G., Spencer Jr, B.F., Jung, H.J., Carlson, J.D.: Dynamic modelling of large-scale magneto rheological damper systems for civil engineering applications. J. Eng. Mech. **130**(9), 1107–1114 (2004)
12. Ogata, K.: Modern Control Engineering, pp. 682–700. Prentice-Hall, New Jersey (1990)
13. Dyke, S.F., Spencer Jr. B.F.: A comparison of semi-active control strategies for the MR damper. In: Proceedings Intelligent Information Systems. IIS'97 no. 6, p. 580 (1997)
14. Chopra, A.K.: Free vibration tests. In: Dynamics of Structures: Theory and Applications to Earthquake Engineering, 2nd edn., pp. 54–57. Prentice Hall, Upper Saddle River, NJ (2001)
15. http://ngawest2.berkeley.edu/

Chapter 20
Study on Treatment of Domestic Wastewater by Modified Bio-Rack System

S. M. Sathe and G. R. Munavalli

Abstract Bio-rack system is an innovative approach to treat domestic wastewater by addressing the problems observed in constructed wetlands. It is rack of vertical pipes in which wetland plants are vegetated. In the present study, the modifications to conventional bio-rack system in terms of multi-species vegetation and different flow pattern are incorporated. Racks were made from waste (packaged drinking water) bottles and vegetated with *Canna Indica* and *Typha Angustata*. Modified and conventional bio-rack systems were operated in continuous mode for 12 and 24 h of hydraulic retention time (HRT) with organic loading rate of 140–300 gm COD/m^2 day. COD removal efficiency for control and modified system for 12 and 24 h HRT was observed to be 55–67% and 65–73%, respectively. Dissolved oxygen level was also higher for modified system than conventional system for both HRTs. The modified system was found to be more efficient than conventional system. The study indicated that modified system showed better performance for COD, BOD and TKN removal and in terms of oxygenation. It can be a low-cost option for the domestic wastewater treatment in decentralized systems.

Keywords Bio-rack wetland · Decentralized treatment · Low-cost treatment · Multi-species vegetation

20.1 Introduction

Domestic wastewater mainly comprises of liquid waste arising from the human habitats. Safe, economic, and effective treatment of sewage is one of the most challenging

S. M. Sathe (✉)
Civil Environmental Engineering, Walchand College of Engineering, 416415 Sangli, Maharashtra, India
e-mail: shreeniwas22.11@gmail.com

G. R. Munavalli
Department of Civil Engineering, Walchand College of Engineering, 416415 Sangli, Maharashtra, India
e-mail: guru.munavalli@walchandsangli.ac.in

© Springer Nature Singapore Pte Ltd. 2020
V. Sivasubramanian and S. Subramanian (eds.), *Global Challenges in Energy and Environment*, Lecture Notes on Multidisciplinary Industrial Engineering, https://doi.org/10.1007/978-981-13-9213-9_20

problems faced worldwide [1]. Globally, billions of people are suffering due to inappropriate sanitation and wastewater treatment. The situation is particularly grave in smaller towns (or peri-urban areas) and rural communities in developing countries [2]. The treatment of wastewater can be provided by centralized or decentralized manner. For urban areas, centralized treatment is beneficial but for rural areas and isolated households, decentralized treatment is suitable. In the absence of sewer lines, decentralized wastewater treatments are cost-effective and convenient way of treating domestic wastewater. Septic tank followed by soak pit is the most commonly used decentralized treatment in India. Problems with safe disposal of septic tank effluents are common where soils are either poorly permeable, making them susceptible to hydraulic failure or highly permeable, allowing rapid by-pass flow [3]. Constructed wetlands have been tried earlier for decentralized treatment of domestic wastewater as a low-cost solution. Various modifications have also been tried in constructed wetlands such as multi-species vegetation, multimedia arrangement, external aeration, and hybrid arrangement [4–10 and 11]. In order to overcome clogging and larger area requirement problems posed by constructed wetlands, bio-rack system has been tried to treat domestic wastewater and low-concentration polluted river water [1, 12–15, and 16].

The study was aimed to modify the conventional bio-rack system and examine its effect on the removal efficiency of pollutants. The modifications done in the existing system include multi-species vegetation and changing flow pattern.

20.2 Materials and Methods

20.2.1 Characteristics of Wastewater

Influent wastewater fed to the system was college hostel wastewater which showed similar characteristics to that of domestic wastewater. The initial characteristics of wastewater during the study period were as follows pH 7.26–7.65, EC 2.42–2.9 mS/cm, TDS 1142–1477 mg/l, BOD 162–186 mg/l, COD 238–264 mg/l, and TKN 46.24–79.45 mg/l.

20.2.2 Experimental Setup

A small-scale pilot reactor of bio-rack system was operated from October 2017 to January 2018 in Walchand College of Engineering, Sangli (Fig. 20.1). Two-reactors control system and modified system consist of two major components (Fig. 20.2):

1. Settling tank: Storage capacity of 180 l (160 l wastewater and 20 l sludge storage).
2. Reactor: Both reactors were circular shaped (0.6 m diameter) with 160 l capacities. The detailed configurations of control and modified system are shown in

Fig. 20.1 Photographic view of pilot scale setup

Table 20.1. The pipes in control system consist of 43 waste (packaged drinking water) plastic bottles arranged as rack having 8 cm diameter. All pipes contained numerous perforations for better liquid transport. Overall flow pattern in control reactor is upflow (Fig. 20.3). The height of vertical pipes is around 600 mm. While in the modified system, the flow pattern was alternate upflow and downflow. Flow pattern was altered by making perforations at different level on concentric circular compartments (Fig. 20.2). Both systems were planted with an equal number of *canna indica* and *Typha angustata.*

20.2.3 Vegetation

Both *Canna Indica* and *T. Angustata* plants were locally collected. Plant roots were cleaned by water. Experiments were performed after acclimatization of plants in

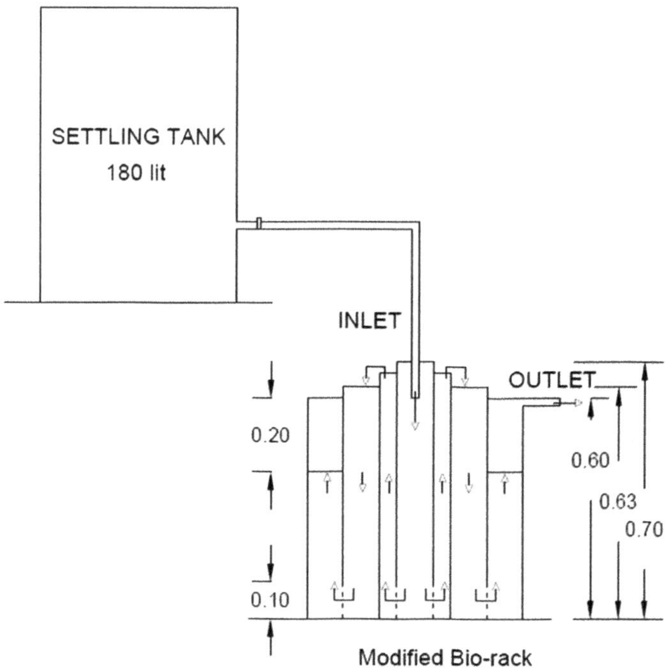

Fig. 20.2 Modified bio-rack reactor configuration

Table 20.1 Experimental reactor configuration

Details	Control system	Modified system
Shape of reactor	Circular	Circular
Diameter of reactor	0.6 m	0.6 m
Volume of reactor	0.17 m^3	0.17 m^3
Initial vegetation	Total-90 plants *Canna Indica*: 40 *T. Angustata*: 50	Total-90 plants *Canna Indica*: 40 *T. Angustata*: 50
Flow pattern	Upflow	Series of downflow and upflow
Details of racks	Plastic bottle pipe with 8 cm diameter and 43 no	4 circular reactors- 10 and 20 cm—PVC pipes 40 cm—Plastic bottles 8 cm diameter, 12 no 60–20 cm length 10 cm diameter, 12 no

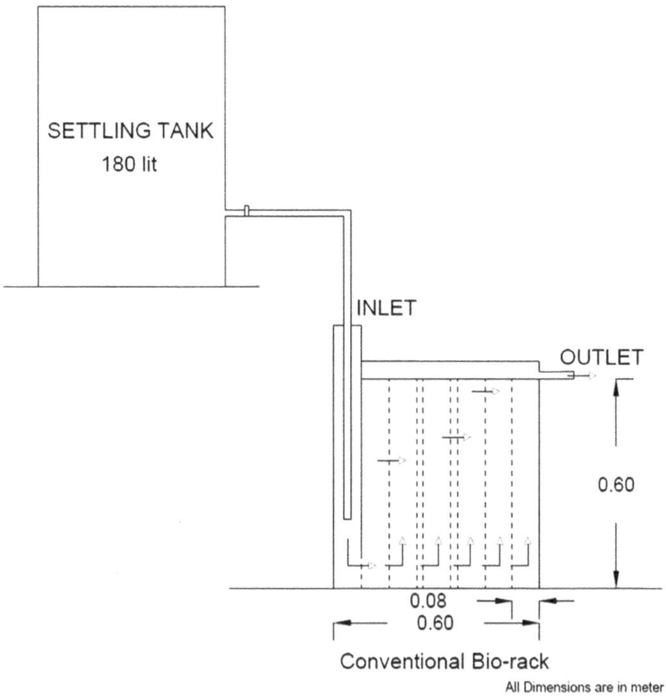

Fig. 20.3 Control bio-rack reactor configuration

wastewater from gradual 25–100% concentration. Length of plant roots in both horizontal and vertical direction was measured before commencement of experiment. The average plant height of *Canna Indica* and *T. Angustata* was 30–40 cm and 40–55 cm, respectively.

20.2.4 Sampling and Analysis

Grab samples of raw wastewater were collected for initial characterization. Wastewater was collected and analyzed from inlet and outlet during each cycle. Mode of operation was continuous and multiple grab samples were collected from the outlet of reactor for each cycle so that representative results for HRT can be obtained. Inlet and outlet samples were analyzed for chemical oxygen demand (COD), biochemical oxygen demand (BOD), and total kjeldahl nitrogen (TKN). Methods for analysis were implemented as per [17].

Flow-through system was monitored and regulated by regulating valve for maintaining the required HRT (12 and 24 h) for system. Desired flow rate was maintained so as to have required organic loading rate (OLR), and hydraulic loading rate (HLR) is mentioned in Table 20.2.

Table 20.2 Experimental conditions for different HRT

Hydraulic retention time (HRT) h	Hydraulic loading rate (HLR) m^3/m^2 day	Organic loading rate (OLR) gm COD/m^2 day	Organic loading rate (OLR) gm BOD/m^2 day
12	1.135	295	200.85
24	0.567	144	98.16

Table 20.3 Root depth and spread

Vegetation	*Canna Indica* (cm)		*T. Angustata* (cm)	
	Length	Spread	Length	Spread
Initial	8.0	8.0	8.0	7.0
Final	19.5	9.0	32.0	8.5

20.3 Results and Discussions

Experiments were carried on hydraulic loading rate and organic loading rate by varying hydraulic retention time for control and modified systems.

20.3.1 Vegetation Growth

Growth of *Canna Indica* and *T. Angustata* was monitored based on the root depth. Root depth before and after experiment was mentioned in Table 20.3.

Both plant species grow well in control and modified system in terms of root depth as well as plant height. It shows that bio-rack supports plants well and the roots can easily penetrate deep inside in the absence of media or supporting matrix. *T. Angustata* roots grow deep as compared to *Canna Indica* in the both systems. It is possible that due to placement of plants in bottles, the horizontal spread is obstructed.

20.3.2 Performance Evaluation for Both Systems

Figures 20.4 and 20.5 show the treatment efficiency in terms of COD and BOD. Performance of modified system was better than control system in terms of COD and BOD removal. Removal efficiency of COD and BOD was achieved for both systems operated at 12 and 24 h HRT. In the bio-rack system, in the absence of media, root growth is governing parameter for attached growth process. Modified system outperformed control system in terms of COD and BOD removal by 5.5–9% and 8.2–9%, respectively. For modified system, due to change in flow pattern, the

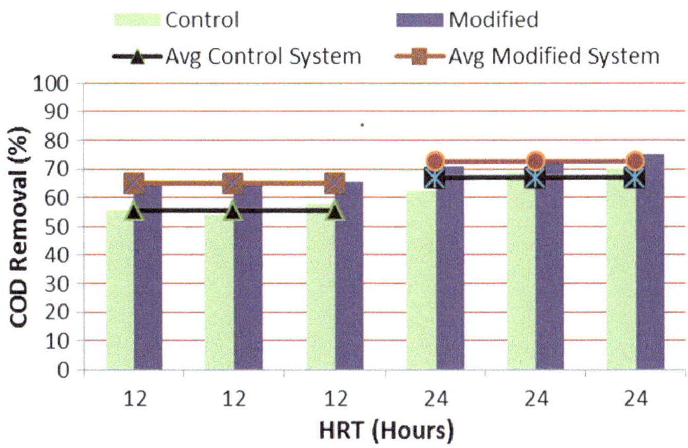

Fig. 20.4 Average COD removal efficiencies for both systems at 12 and 24 h HRT

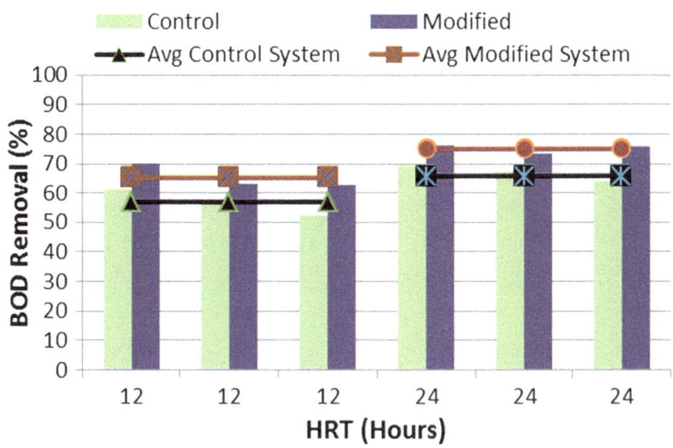

Fig. 20.5 Average BOD removal efficiencies for both systems at 12 and 24 h HRT

contact time of wastewater with root matrix is increased which may have resulted in higher removal efficiency.

Figure 20.6 shows the total kheldahl nitrogen removal efficiency of both systems. The results indicated that performance of modified system was better than control system. Total kheldahl nitrogen removal efficiency was comparatively less for both systems. Nitrification occurs under aerobic condition while denitrification occurs under anoxic condition. DO level in both systems seem to have exceeded required anoxic condition (i.e., 0.2–0.5 mg/L) [18]. Due to higher DO levels, denitrification could have not achieved which may have affected the total nitrogen removal efficiency.

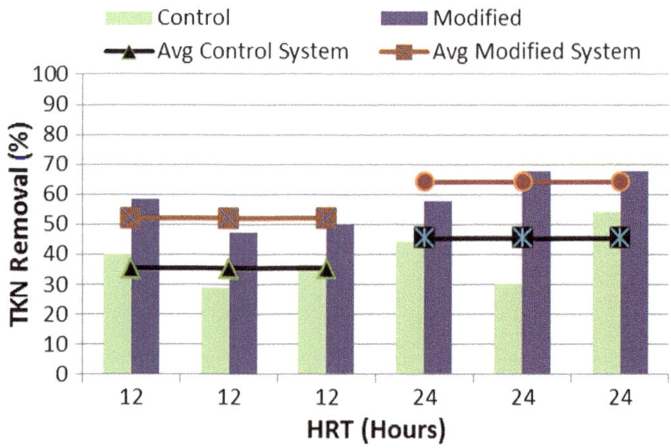

Fig. 20.6 Average TKN removal efficiencies for both systems at 12 and 24 h HRT

20.3.3 Dissolved Oxygen Concentration in Both Systems

As shown in Fig. 20.7, the dissolved oxygen concentration in the effluent is observed to be higher in modified system. Despite the absence of DO in influent, increase in the dissolved oxygen level was seen for both systems. DO concentration for modified system was 0.6–1 mg/l higher than control system showing that more roots came in contact with wastewater during its passage. Due to higher DO, higher aerobic conditions were maintained in modified system which also increases the removal rate for organic matter.

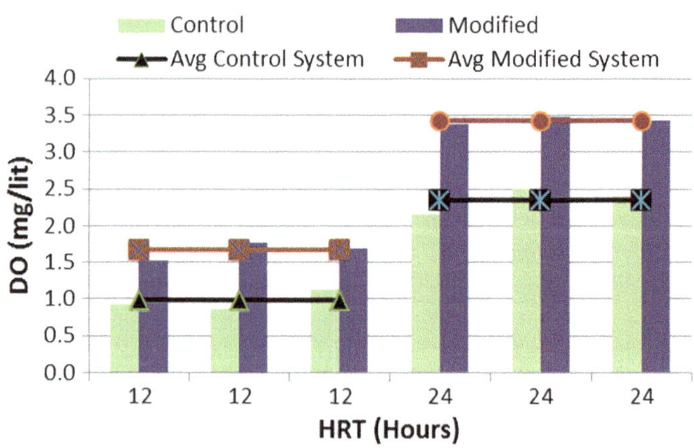

Fig. 20.7 Average DO concentration in the effluent for both systems at 12 and 24 h HRT

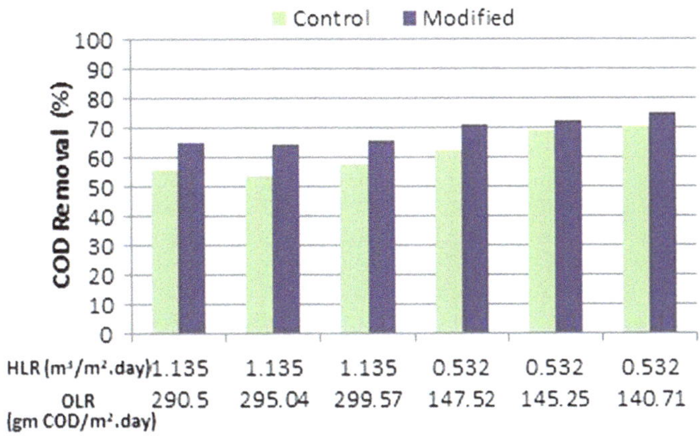

Fig. 20.8 Effect of HLR and OLR on COD removal efficiency

20.3.4 Effect of HLR and OLR on COD Removal Efficiency

The relationship of HLR and OLR with COD removal efficiency is given in Fig. 20.8.

Results demonstrate that with an increase in HLR from 0.532 to 1.135 m/day, the removal efficiency of COD in both systems decreased. Increase in HLR reduces the contact time of wastewater with microbes and plant roots; hence, it reduces the oxygen level which leads to lesser aerobic conditions and hence the COD removal efficiency was also reduced. Treatment of wastewater in the system can take place aerobically as well as anaerobically but due to higher rate in aerobic treatment, it is preferred over anaerobic condition.

20.4 Conclusion

This study assessed the potential benefits of modified bio-rack system in treating the domestic wastewater over conventional system. In both systems, vegetation grows well in terms of height and root penetration which shows that bio-rack supports vegetation. Of the two vegetation species, *T. Angustata* roots development is better than *Canna Indica*. Modifications are done in terms of multi-species vegetation and changing flow pattern results in higher COD, BOD, and TKN removal than control system by 5.5–9%, 8.2–9%, and 15–17%, respectively. The significant DO increase is achieved by bio-rack system. HLR affects the performance of bio-rack system where increase in HLR reduces the performance efficiency. Modified system can well be a low-cost option for domestic wastewater treatment.

References

1. Valipour, A., Raman, V.K., Ghole, V.S.: A new approach in wetland systems for domestic wastewater treatment using *Phragmites sp*. Ecol. Eng. **35**, 1797–1803 (2009)
2. Singh, N., Kazmi, A., Starkl, M.: A review on full-scale decentralized wastewater treatment systems: techno-economical approach. Water Sci. Technol. **71**(4), 468–478 (2015)
3. Tanner, C., Sukias, J., Headley, T., Yates, C., Stott, R.: Constructed wetlands and denitrifying bioreactors for on-site and decentralized wastewater treatment: comparison of five alternative configurations. Ecol. Eng. **42**, 112–123 (2012)
4. Coleman, J., Hench, K., Garbutt, K., Sexstone, A., Bissonnette, G., Skousen, J.: Treatment of domestic wastewater by three plant species in constructed wetlands. Water Air Soil Pollut. **128**(3–4), 283–295 (2001)
5. Abidi, S., Kallali, H., Jedidi, N., Bouzaiane, O., Hassen, A.: Comparative pilot study of the performances of two constructed wetland wastewater treatment hybrid systems. Desalination **246**, 370–377 (2009)
6. Babatunde, A., Zhao, Y., Zhao, X.: Alum sludge-based constructed wetland system for enhanced removal of P and OM from wastewater: concept, design and performance analysis. Biores. Technol. **101**, 6576–6579 (2010)
7. Wu, S., Austin, D., Liu, L., Dong, R.: Performance of integrated household constructed wetland for domestic wastewater treatment in rural areas. Ecol. Eng. **37**, 948–954 (2011)
8. Abou-Elela, S., Hellal, M.: Municipal wastewater treatment using vertical flow constructed wetlands planted with *Canna*, *Phragmites* and *Cyprus*. Ecol. Eng. **47**, 209–213 (2012)
9. Avila, C., Garfi, M., Garcia, J.: Three-stage hybrid constructed wetland system for wastewater treatment and reuse in warm climate regions. Ecol. Eng. **61**, 43–49 (2013)
10. Vymazal, J., Kropfelova, L.: Multistage hybrid constructed wetland for enhanced removal of nitrogen. Ecol. Eng. **84**, 202–208 (2015)
11. Fountoulakis, M., Daskalakis, G., Papadaki, A., Kalogerakis, N., Manios, T.: Use of halophytes in pilot-scale horizontal flow constructed wetland treating domestic wastewater. Environ. Sci. Pollut. Res. **24**, 16682–16689 (2017)
12. Valipour, A., Raman, V.K., Ghole, V.S.: Application of patent Bio-rack wetland system using *Phragmites sp.* for domestic wastewater treatment in the presence of high Total Dissolved Solids (TDS) and heavy metal salts. J. Environ. Sci. Eng. **53**(3), 281–288 (2011)
13. Valipour, A., Hamnabard, N., Woo, K., Ahn, Y.: Performance of high-rate constructed phytoremediation process with attached growth for domestic wastewater treatment: effect of high TDS and Cu. J. Environ. Manage. **145**, 1–8 (2014)
14. Wang, J., Zhang, L., Lu, S., Jin, X., Gan, S.: Contaminant removal from low-concentration polluted river water by the bio-rack wetlands. J. Environ. Sci. **24**(6), 1006–1013 (2012)
15. Wang, J., Ying, Z., Shaoyong, L., Can J., Ganshu.: A comparison of two aquatic plants purifying the low-concentration polluted river water in the bio-rack wetland. In: Proceedings of World Automation Congress (WAC) (2012)
16. Jamshidi, S., Akbarzadeh, A., Woo, K., Valipour, A.: Wastewater treatment using integrated anaerobic baffled reactor and Bio-rack wetland planted with *Phragmites sp.* and *Typha sp.* J. Environ. Health Sci. Eng. **12**, 131 (2014)
17. APHA.: Standard methods for examination of water and wastewater. 22nd edn. American Public Health Association, Washington, DC (2012)
18. Dong, H., Qiang, Z., Li, T., Jin, H., Chen, W.: Effect of artificial aeration on the performance of vertical-flow constructed wetland treating heavily polluted river water. J. Environ. Sci. **24**(4), 596–601 (2012)

Printed by Printforce, the Netherlands